Political Agroecology

Advancing the Transition to Sustainable Food Systems

Advances in Agroecology

Series Editors:

Clive A. Edwards, The Ohio State University, Columbus, Ohio
Stephen R. Gliessman, University of California, Santa Cruz, California

For more information about this series, please visit: www.crcpress.com/Advances-in-Agroecology/book-series/CRCADVAGROECO

Political Agroecology

Advancing the Transition to Sustainable Food Systems

Manuel González de Molina
Paulo Frederico Petersen
Francisco Garrido Peña
Francisco Roberto Caporal

CRC Press
Taylor & Francis Group
Boca Raton London New York

CRC Press is an imprint of the
Taylor & Francis Group, an informa business

CRC Press
Taylor & Francis Group
52 Vanderbilt Avenue,
New York, NY 10017

International Standard Book Number-13: 978-1-138-36922-1 (Paperback)
International Standard Book Number-13: 978-1-138-36923-8 (Hardback)

Visit the Taylor & Francis Web site at
www.taylorandfrancis.com

and the CRC Press Web site at
www.crcpress.com

Cover Art by Lucia Vignoli

The cover illustration, "Buen Vivir," was created by Lucia Vignoli and Joana Lyra, artist-faculty members of the ArteGestoAção Research Group (ArtGestureAction) linked to the Instituto Nacional de Educación de Sordos [National Institute of Deaf People], Brazil. Produced with manual paintings and stamps, the image is inspired by a traditional West African communication system, the Adinkras stamps, vehicles of information and values with aesthetic, ethical and political meanings. The composition reveals a system of interactions in which humans and non-humans are organically integrated into the reproduction of "Buen Vivir".

Contents

Chapter 6   The Agents of the Agroecological Transition 119

   6.1   "Food Populism": Building Social Majorities of Change 119
   6.2   Peasants: Central Actors in the Agroecological Transition 123
   6.3   Peasant Conditions under Capitalism and Agricultural
         Industrialization...128
   6.4   The "New Peasants" 131
   6.5   Agroecology and Feminism: The Central Role of Women 133
   6.6   Politicizing Food..135
   6.7   Agroecological Movements as "New Green" Movements......138

Chapter 7   The Role of the State and Public Policies 143

   7.1   Public Policies from a Political Agroecology Perspective 144
   7.2   Experiences in Public Policies Favoring Agroecology...........146
   7.3   Main Conclusions of the Analysis of Public Policies..............151
   7.4   Public Policies to Scale Up Agroecology 153
   7.5   An Agroecological Approach to the Design and
         Implementation of Public Policies..154
   7.6   Public Policies that Lead to Scaling Up 157
         7.6.1   Program for the Construction of Cisterns, Brazilian
                 Semiarid Region...158
         7.6.2   Organic Agriculture Program in Cuba 159
         7.6.3   Organic Foods for Social Consumption in
                 Andalusia, Spain 159
         7.6.4   Biofertilization and Biological Control Input
                 Programs in Cuba...160
         7.6.5   The National Policy of Technical Assistance and
                 Rural Extension, Brazil..160
         7.6.6   Institutional Purchasing, Brazil...................................161
         7.6.7   Olive "Residues" Composting Program in
                 Andalusia, Spain 161
         7.6.8   ProHuerta's Program, Argentina.................................162
         7.6.9   The National Policy of Agroecology and Organic
                 Production, Brazil 162
         7.6.10 State Policy on Organic Farming (2004) and
                 Organic Mission (2010), Sikkim, India 163
         7.6.11 Milan Urban Food Policy Pact (2015) 164
         7.6.12 Advances in Professional Training and Support for
                 the Organization of Agroecology Hubs in Brazilian
                 Educational Institutions 164

References 167
Index 195

About the Authors

Manuel González de Molina

Full Professor of Modern History. Coordinator of the Agro-Ecosystems History Laboratory (Universidad Pablo de Olavide, Seville, Spain). Co-director of Master Degree Program on Agroecology at International University of Andalusia from 1996 to present. President of the Spanish Society for Agrarian History (www.seha.info) and member of the editorial board of the ISI-refereed journal *Historia Agraria* [*Agrarian History Review*], *Anthropoce* and *Sustainability*. Vice-President of the Spanish Society for Organic Agriculture (SEAE) from 2006 to 2014. Minister of the Department of Organic Agriculture of the Andalusia Government (Spain) from 2004 to 2007. Author of several books, among the most recent: *The Social Metabolism. A Socioecological Theory of Historical Change* (Springer, 2014), *Energy in Agroecosystems: A Tool for Assessing Sustainability* (CRC Press, USA, 2017). Author of 100 articles published in journals such as *Environment and History, Ecological Economics, Land Use Policy, Environmental History, Regional Environmental Change, Ecology and Society, Agroecology and Sustainable Food Systems, Journal of Interdisciplinary History*, among others.

Paulo Frederico Petersen

Doctorate in Environmental Studies from Pablo de Olavide University (Spain), Master in Agroecology and Rural Development from the Universidad Internacional de Andalucía (Spain), graduated in Agronomy from the Federal University of Viçosa (Brazil). Executive coordinator of the NGO AS-PTA – Family Farming and Agroecology, former President (from 2013 to 2014) and current Vice-President of ABA-Agroecologia (Brazilian Association of Agroecology). Editor-in-chief of *Agriculturas: experiencias em agroecologia* magazine. Member of the National Commission of Agroecology and Organic Production (CNAPO), and member of its Coordinating Table from 2014 to 2018. Member of the Editorial Board of *Agroecology and Sustainable Food Systems* (USA), *Revista de Agroecología* (Spain), *Revista Brasileira de Agroecologia* (Brazil), and the *Coleção Transição Agroecológica* published by Brazilian Agricultural Research Enterprise (Embrapa).

Francisco Garrido Peña

PhD in Philosophy of Law (Universidad de Granada) and full professor of Politics and Law at Universidad de Jaén (Spain). His research activity is focused on Political Ecology, Ethics, and Institutional Design. In Political Ecology, the topics of research have been the political ecology of time (intertemporal discounts), political sovereignty, ecological theory of the State and political power, ecological conflicts, social movements, gender and ecology, animalism and Political Agroecology. In the discipline of ethics the main topics have been: ecological ethics, zooethics, public bioethics, experimental ethics, altruism, intergenerational solidarity, sense of justice. And in the field of institutional design: design of intergenerational institutions, institutions for sustainability, collective decisions in public health and the environment, citizen participation. These academic activities have been complemented with his social and political activism in the environmental movements. He was elected Deputy of the Greens in the parliament of the region of Andalucía between 1994 and 1996. He was also Deputy of the Greens in the Spanish parliament between 2004 and 2008. He was Spanish confederal speaker of the Green Party (Los Verdes) between 1991 and 1994 and 2004 and 2008.

Francisco Roberto Caporal

PhD in Agroecology, Peasantry and History, from the Institute of Sociology and Peasant Studies (Córdoba University, Spain). Master in Rural Extension by Federal University of Santa Maria (1991) and graduate in Agronomy, Federal University of Santa Maria (1975). Currently, he works as Associate Professor of the Federal Rural University of Pernambuco, in the Department of Education, teaching in Rural Extension. Member of the Agroecology and Peasantry Nucleus – NAC/UFRPE. He was Technical Director of EMATER-RS, from 1999 to 2002, as Deputy Director of the Department of Technical Assistance and Rural Extension (DATER) and General Coordinator of Ater and Education, in the same Department of the Secretariat of Family Agriculture, Agricultural Development, from 2004 to 2010 (Agrarian Development Minister of Brazil). He has experience in Agronomy, with emphasis on Technical Assistance and Rural Extension, working mainly in the following subjects: technical assistance and rural extension, sustainable rural development, Agroecology, environment and family agriculture, training of rural extension agents. President of Associação Brasileira de Agroecologia (ABA-Agroecologia) from 2006 to 2010.

Introduction

From a biophysical point of view, the industrialization of agriculture led to a fundamental reconfiguration of agricultural production and the way in which our endosomatic consumption is satisfied. The possibility of injecting large amounts of energy and materials offered by oil and its associated technologies radically changed the world's agricultural scenario during the twentieth century. Basic functions that in past times had been fulfilled by the land (production of traction power, fuels, fiber, feed, basic foodstuffs for human consumption, etc.), and to which a fairly large portion of territory was dedicated, have disappeared, giving rise to a specialized landscape, peppered with constructions and areas used for urban-industrial properties (Agnoletti, 2006; Guzmán Casado & González de Molina, 2009; González de Molina & Toledo, 2014). From 1961 to 2016, the world's production of cereals multiplied almost by four, surpassing the corresponding growth rate in global population, and the availability of cereals per capita rose by 60% as a result (www.fao.org/faostat, accessed 15 March 2019).

Despite this huge productive effort, rural poverty, hunger, and endemic malnutrition continue to exist. The Food and Agriculture Organization has estimated that over 825 million people in the world are victims of famine or malnutrition (see www.fao.org). The dominant food regime is incapable today of nourishing the whole of humankind – it doesn't really seem to be oriented toward that goal either – even though there is enough raw harvest to do so. Scant progress has been made to eradicate rural poverty. Apparently, agriculture and the food system continue to supply endosomatic energy, which ensures nourishment and the reproduction of the human species, but their role is currently experiencing major changes (Francis et al., 2003). From being a vital provider of energy, agriculture has become a consumer of energy. Deprived of its external energy subsidy, it might not work any longer (Leach, 1976; Pimentel & Pimentel, 1979; Gliessman, 1998; Guzmán Casado & González de Molina, 2017). Agriculture has become a subsidiary activity within industrial economies, being valued mainly as a food and raw materials supplier and to a much lesser extent as a provider of other goods and services, e.g., environmental ones. The biomass produced is but one among all the material flows, and its weight is losing ground to global social metabolism (Krausmann et al., 2017a).

The food market has gone global, forcing agricultural products to travel long distances before reaching consumers' tables; even when eaten fresh, they require huge logistical infrastructures. Processed food has overtaken fresh food and the amount of meals eaten out increases day by day. Furthermore, sophisticated new appliances, using gas or electricity, take part in human nutrition and have increased food's energy cost (Infante et al., 2018). New activities, unheard of in the past, intervening between production and consumption, have come to the fore and have become paramount: food transformation and delivery. The way food is eaten nowadays causes not only huge health problems to humans but also creates unhealthy agricultural systems, including those in third countries. There is a long story behind each piece of food we eat: a long chain of links that multiplies energy and material consumption together with pollutant emissions and unbalanced trade. Thus, food consumption is turned into a process rife with social and environmental impacts. As a result, the distribution of food in the world has become increasingly unfair. On the one hand, a large portion of the world's population cannot reach a minimum consumption of calories and nutrients to the extent that famine and undernourishment have become structural phenomena. On the other hand, a similar portion of the world's population is overfed and suffers severe health problems which financially strain their health systems (IPES-Food, 2016).

These are manifestations of a crisis arising from mechanisms that lie at the very heart of the dominant food regime. Allowing them to continue leads to accelerating social and environmental deterioration and, if not remedied, the system is likely to collapse. The only viable solution lies in building a markedly different food system, based on sustainable forms of production and processing, distribution, and consumption. The multitude of agroecological experiences that exists the world over prefigures and constitutes the basis of this alternative food regime. The challenge is to scale up both horizontally (scaling out) and vertically (scaling up). Indeed, the majority of agroecological experiences, linked to social movements, non-governmental organizations, academic institutions and, to a significantly lesser extent, governments, remain widely restricted to farm or community experiences, encouraging research, participatory action and design of sustainable rural development strategies or urban food supply. To scale out and up, it is essential to change the institutional framework that is currently maintaining, despite its unfeasibility, today's corporate food regime (CFR). This change must necessarily be political. However, agroecological movements are characterized by the scarcity of political proposals that reach beyond the local sphere. This is because the nexus between Agroecology and politics is not fully perceived as a fundamental link for developing and maintaining agroecological experiences and, above all, to generalize them. So Agroecology is not entirely prepared to face this challenge. The aim of this book is to build a political theory that makes the scaling-up of agroecological experiences possible, turning them into the foundation of a new and alternative food regime.

The link between politics and Agroecology is not new. Many authors have emphasized the need for socioeconomic structural reforms in order to achieve sustainable agrarian systems (Gliessman, 1998; Rosset, 2003; Levins, 2006; Holt-Giménez, 2006; Perfecto et al., 2009; Altieri & Toledo, 2011; Rosset & Altieri, 2017; Giraldo, 2018). However, this concern is still far from widespread and has not been fully internalized by agroecologists and agroecological movements. Meanwhile, some academic and institutional sectors of Agroecology, that foster a purely "technical" vision of Agroecology, are becoming increasingly influential. They promote technological solutions rather than institutional or social change solutions to the unsustainability problems produced by the CFR. Ignoring politics or relegating it to a secondary place results, on the one hand, in agroecological experiences lacking efficiency and stability and barely reaching the required size, thus hindering the necessary process of scaling out and up; and on the other hand, they result in spreading the false idea that technological innovation alone, without substantial social and economic change, will achieve more sustainable food systems. The first leads to inefficiency, the second to inactivity, and both sever the possibilities of Agroecology becoming an alternative to the current food regime.

According to Wezel et al. (2009), most of the agroecological literature is devoted to the technical–agronomic aspects of agroecosystem management. However, some authors have been stressing the need to broaden the scope beyond production, so that Agroecology is effectively consolidated as a guiding approach leading agriculture toward sustainability (Francis et al., 2003; Altieri & Toledo, 2011; Gliessman, 2015; Pimbert, 2015). Faced with the temptation to consider Agroecology as the sum of three types of action, as science, as practice, and as a social movement, considered even in an independent way, theory and practice must be understood in an indissoluble way. In fact, Agroecology has been synergistically combining these three dimensions of action, condensing its analytical approach, its operational capacity and its political incidence into an indivisible whole. Agroecology is therefore a "transformative science" (Levidow et al. 2014; Schneidewind et al., 2016), that is, an analytical approach that incorporates a critique of the governance mechanisms adopted in the CFR (McMichael, 2009; Holt-Giménez & Altieri, 2013) and, on the basis of transdisciplinary and participatory perspectives, is used to design and construct social dynamics of change, according to principles of socioecological sustainability (Méndez et al., 2013).

Indeed, we will attempt to demonstrate in this book that the food regime's crisis does not derive solely from its environmental impacts: the roots of the crisis lie in the institutional rules regulating and governing the current regime. The causes should not be confused with the consequences. Therefore, Agroecology cannot limit itself to pointing out unsustainable factors in agroecosystems, proposing management approaches and practices that will restore the sustainability of these factors is essential. As stated by Gliessman (2011, 347), Agroecology consists above all of powerful instrument to achieve change in the food regime, in other words, a massive redesign of the economic structures that govern it. This practical dimension of Agroecology requires politics, that is, the disciplines responsible for designing and implementing institutions that make food systems' sustainability possible. The search for sustainability implies a transformation of the dynamics of agroecosystems that can only be achieved by social agents and their institutional mediation. Agroecology, however, is not yet equipped with the analytical instruments and criteria required to define strategies that could guide this change. Politics must develop within the heart of Agroecology to provide agroecologists with instruments for analysis and sociopolitical intervention that would allow them to go beyond local experiences, encouraging their generalization and the essential changes in the food system at a higher territorial scale (scaling up). This book is devoted to this necessary political dimension of Agroecology; it proposes theoretical and epistemological foundations of a new theoretical and practical field of work for agroecologists: *Political Agroecology*.

The purpose of this book is to establish a common framework of analysis of agroecological collective action. Epistemological and theoretical arguments are provided, pushing toward the development of this field of Agroecology and making the fight for food sustainability operative. The objective is also to create a food crisis narrative that can serve as a common framework to guide collective agroecological action. In short, this book aims to lay the theoretical and methodological foundations of a common agroecological strategy, covering the different levels of collective action and the different instruments with which it can be developed.

The first chapter is dedicated to the theoretical grounding of the key role that institutions and, therefore, politics have in the dynamics of agroecosystems and food regimes as regulators of the energy, materials and information flows that they exchange with their physical and social environment. The second chapter describes the paths followed by agricultural industrialization and highlights the main components of the institutional framework responsible for this process. Major attention is given to describing and analyzing the emergence and hegemonic development of the food regimes that have been governing the world since the end of the nineteenth century. The third chapter is dedicated to characterizing, from a sociometabolic and institutional point of view, the current crisis of the world food system, governed by the CFR, and the risks of collapse to which its hegemony leads. The fourth chapter begins the second most propositive part of this book, which discusses principles and criteria for the design of institutions and the cognitive frameworks that Political Agroecology must provide in order to promote a new sustainable food regime. The chapter also discusses the most effective collective action to promote food change, an action that cannot be segmented into agroecosystem management practices, transformative social movement practices, or institutional action: multilevel collective action that strategically combines all these dimensions is called for. The fifth chapter discusses the best strategy for agroecological experiences to scale out and scale up, constituting a viable alternative to the CFR. Based on this strategy, tasks are defined to be undertaken both by the agroecological movements and for the building of local-scale agroecological-based food systems. Chapter 6 addresses the major players of food change. It is imperative that peasants and new peasants be given a key role, but not only them. This chapter stresses the need for women and consumers to play a leading role. Achieving social majorities of change will be difficult without their contribution. To cement this alliance, we propose that the vast majority of citizens be organized and mobilized around demands for food sovereignty, that is, based on *food populism*. Finally, Chapter 7 analyzes the experiences of public policies to date that have favored Agroecology and, drawing lessons from these practices, proposes policies that contribute to the upscaling of Agroecology.

The authors would like to thank Stephen Gliessman for both his intellectual inspiration and his efforts to make this book possible. We also owe a debt of gratitude to Miguel Altieri, Clara Nicholls, and Ernesto Méndez for their important recommendations regarding the writing of this book. We would like to extend our thanks to Víctor Toledo, Jan Douwe van der Ploeg, Jaime Morales, Eric Holt-Giménez, Peter Rosset, Gloria Guzmán, David Soto, Daniel López García, Ángel Calle, Eduardo Sevilla Guzmán, José Antônio Costabeber (in memoriam), the Nucleus of Agroecology and Peasantry members of the UFRPE, Claudia Schmitt, Eric Sabourin, Silvio Gomes de Almeida and other companions from AS-PTA, colleagues with whom we have collaborated toward an environmentally healthy, economically viable, socially just and politically democratic food system. Without their inspirations this book would not have been possible. Finally, we would like to thank Christine Sagar for her help in the translation and linguistic revision of the manuscript, which has clearly improved it.

1 Theoretical Foundations of Political Agroecology

Agroecosystems are artificialized ecosystems that shape a particular subsystem operating within the general metabolic flows between society and nature; therefore, they are a product of the socioecological relations. For example, using or changing crops is a decision that often has socioeconomic roots and, at the same time, environmental consequences. These kinds of socioecological relations are part of social relations generally, in which power and conflict are present. Consequently, even in the simplest societies, technologically speaking, the specific assembly of each agroecosystem responds to different types of institutions, forms of knowledge, world views, rules, norms and agreements, technological knowledge, means of communication and governance, and forms of ownership (González de Molina & Toledo, 2011, 2014). An agroecosystem's sustainability does not result solely from a series of physical and biological properties: it also reflects power relations. Agroecology therefore needs to be placed within a political framework.

In this regard, the quest for sustainable agricultural ecosystems requires *Political Agroecology*, which is a new way of organizing agroecosystems and agricultural metabolism in general. In the same way that political power articulates different subsystems in a socioenvironmental system, Political Agroecology should articulate an agroecosystem's different subsystems by organizing energy, material, and information flows. Political Agroecology is tasked with this articulation, programming and functional orientation, bringing continuity and order to the agroecosystem's evolution. This chapter seeks to define Political Agroecology and to develop its epistemological and theoretical basis.

1.1 POLITICAL AGROECOLOGY: A TENTATIVE DEFINITION

Political Agroecology would be the application of Political Ecology to the field of Agroecology, or a close association between these fields (Toledo, 1999; Forsyth, 2008), but there is no agreement as to what Political Ecology actually is (Peterson, 2000; Blaikie, 2008, 766–767). The term gives rise to many meanings and understandings regarding its goal, but all of them have in common a Political Economy approach to natural resources and its preferential application to developing countries (Blaikie, 2008, 767). We share the interpretation of Gezon and Paulson (2005) for whom "the control and use of natural resources, and consequently the course of environmental change" are shaped by "the multifaceted relations of politics and power, and the cultural constructions of the environment". In this sense, Political Ecology combines political and ecological processes when analyzing environmental change and it could also be understood as "the politics of environmental change" (Nigren & Rikoon, 2008, 767). Paraphrasing Blaikie and Brookfield, we could say that "Political Ecology [is] an approach for studying ecological and social change" (Blaikie & Brookfield, 1987), but *together*. In other words, Political Ecology is an approach to studying socioecological change in political terms. Based on Paulson et al. (2003, 209) and Walker (2007, 208), we could say that Political Agroecology should "develop ways to *apply* the methods and findings [of Political Ecology research] in addressing" socioecological change in agroecosystems and the whole food system.

However, Political Agroecology is not only a research subject. It has another closely related practical dimension that is regarded as a core goal: achieving agrarian sustainability. Many agroecologists are involved in a form of " 'popular Political Ecology' that ties research directly to activist efforts to improve human well-being and environmental sustainability through various forms of local, grassroots activism and organization" (Walker, 2007, 364). In this respect, Political Agroecology should branch off into two directions: into an ideology, in competition with others, dedicated to disseminating and turning the organization of ecologically and sustainably based agroecosystems into the dominant system (Garrido, 1993); but also into a disciplinary field responsible for *designing and producing actions, institutions and regulations aimed at achieving agrarian sustainability*.

Political Agroecology is based on the fact that agrarian sustainability cannot be achieved using only technological (agronomical or environmental) measures helping to sustainably redesign agroecosystems. Without profoundly changing the institutional framework in force, it will not be possible to spread successful agroecological experiences nor to effectively combat the ecological crisis. Consequently, Political Agroecology examines the most suitable course of action today and how to best use the instruments that make institutional change possible. Such a change, in a world still organized around nation states, is only possible through political mediation. In democratic systems, for example, it implies collective action through social movements, electoral political participation, alliance games between different social forces to build majorities of change, etc. In other words, it calls for the creation of strategies that are essentially political. The two main objectives of Political Agroecology precisely comprise: the design of institutions (Ostrom, 1990, 2001, 2009) that favor the achievement of agrarian sustainability, and the organization of agroecological movements in such a way that they can be implemented.

Political Agroecology thus goes beyond proposing a specific program. For example, the demand for alimentary sovereignty, promoted by *Vía Campesina* and other social movements is *a* specific proposal for a program that *can* emerge from applying Political Agroecology to the current conditions of the ruling food regime. Political Agroecology is responsible for establishing it and, as a new branch of Agroecology, it is *not* a political proposal or program to achieve agrarian sustainability. Political Agroecology is not a new term for food sovereignty. It seeks to produce knowledge that allows Agroecology and food sovereignty to be put into practice, exploiting the knowledge accumulated by Political Ecology and the experience of social movements.

Political Agroecology thus requires to be grounded in a rigorous socioecological framework that adequately spells out the roles of institutions and the necessary means to establish or change them, anchored in the indissoluble nexus established between human beings and their biophysical environment. In the sections that follow we explore societies' biophysical foundations and draw attention to the determining role of institutional arrangements in their dynamics. This approach is later applied to agroecosystems, understood as the materialization of socioecological relationships in the field of agriculture. We also draw attention to the key role of institutions that regulate their dynamics.

1.2 A THERMODYNAMIC APPROACH TO SOCIETY

Our conception of Political Agroecology is based on a biophysical reading of society, in accordance with its socioecological nature: social systems are subject to the laws of thermodynamics. That means, therefore, that the laws of nature operate on and affect human beings and the devices they build. We thus assume that entropy is common to all natural processes, be they human or other, and it may be the most relevant physical law to explain the evolution over time of the human species. Our understanding of the material structure, functioning, and dynamics of human societies is thus grounded in a thermodynamic understanding, as in the case of biological systems which they also part of.

From a thermodynamic point of view, all human societies share the need for controlled, efficient processing of energy extracted from the surroundings with other physical and biological systems. Such is the proposal of Prigogine (1983) regarding non-equilibrium systems (thermodynamics of irreversible processes), which is one of the basic concepts of our socioecological approach to power and politics: generation of *order out of chaos*. Because the natural trend of societies – as any physical and biological system – is toward a state of maximum entropy, social systems depend on building dissipative structures for balancing this trend and keeping away from maximum entropy (Prigogine, 1947, 1955, 1962). These structures are maintained thanks to the transfer by the system of a part of the energy being dissipated by its conversion processes (Glandsorff & Prigogine, 1971, 288). The transfer takes place by using flows of energy, materials, and information to perform work and dissipate heat, consequently increasing their internal organization. Order emerges from temporal patterns (systems) within a universe that, as a whole, moves slowly toward thermodynamic dissipation (Swanson et al., 1997, 47). Prigogine described this configuration of dissipative structures as a process of self-organization of the system.

Although human societies share the same evolutionary precepts as physical and biological systems, they represent an *innovation* that differentiates them and makes their dynamic specific, adding complexity and connectivity to the whole evolutionary process. Social systems cannot be explained by a simple application of the laws of physics, even though human acts are subject to them. The reason for this is that although evolution is a unified process, human society is an evolutionary innovation emerging from human beings' reflective (self-referring) capacity, which is more developed than in any other species. The most direct consequence of this human mental feature is the capacity – not exclusive among higher-order animals, but rare – for building tools and, therefore, for using energy *outside* the organism, i.e., the use of exosomatic energy. To build and use tools, information and knowledge needed to be generated and transmitted, i.e., the generation of culture was required. Culture involves a symbolic dimension containing, besides knowledge, beliefs, rules and regulations, technologies, etc. Accordingly, evolutionary innovation encompasses human capability regarding the exosomatic use of information, energy, and matter, also giving rise to a new type of complex system: the *reflexive complex system* (Martínez-Alier et al., 1998, 282) or *self-reflexive system* and *self-aware system* (Kay et al., 1999; Ramos-Martin, 2003). This feature will be instrumental because it gives social systems a unique *neopoietic* capacity absent from other systems or species, and that confers an essential, creative dimension to human individuals and – more so – to collective actions.

In analogy to living organisms, culture is the transmission of information by non-genetic means, a metaphor that became popular in the academic world. Cultural evolution has been described as an extension of biological information *by other means* (Sahlins & Service, 1960; Margalef, 1980), and a parallel has been drawn between the diffusion of genes and that of culture. Culture can thus be understood as an innovative manifestation of the adaptive complexity of social systems; it is the name of a new genus of complexity provided by the environment for perpetuating and reorganizing a particular kind of dissipative system: social systems (Tyrtania, 2008, 51). Culture is but an emergent property of human societies. Its performative or neopoietic character, its creative nature (Maturana & Varela, 1980; Rosen, 1985, 2000; Pattee, 1995; Giampietro et al., 2006) enables the configuration of new and more complex dissipative structures at even larger scales by means of technology (Adams, 1988).

As we have seen, organization is an autopoietic product in which flows of information have a definitive influence. There is no structure without information, as has been demonstrated in the biological world (Margalef, 1995). In the social world, systems are also subjected to the laws of thermodynamics, given that they occupy time, space, and energetic resources. Applying thermodynamics in Boltzmann's statistical terms provides an explanation of flows as a unidirectional and irreversible process going from its state of order – its more evident manifestation – to a state of disorder, whose organizational properties have disappeared. Therefore, the main function of these flows is negentropic: "Information, in this technical sense, is the patterning, order, organization, or non-randomness of a system. Shannon showed that information (H) is the negative of entropy (S)" (Swanson et al., 1997, 47). Therefore, flows of information are here considered capable of reordering and reorganizing the different components of the physical, biological, and social systems in which they function. That is, they have characteristics that produce action (change). Information flows are the basic vital ingredients of the processes of organization of social systems. Information is here defined in a pragmatic or operative way as a codified message, which decision-makers can use to regulate levels of entropy.

As H. Gintis (2009, 233) remarked, culture can also be considered as an *epigenetic* mechanism of horizontal, intragenerational transmission of information among humans, i.e., the system's memory along the evolutionary process. In sum, dissipative structures of social systems are designed and organized through culture. There is insufficient space here to present a theory of information flows and their role in complex adaptive systems. Niklas Luhmann (1984, 1998) largely developed this theory, and we direct interested readers to his writings. Luhmann's theory of autopoietic social systems is useful to elaborate a theory of information flows and their role of organizers of the dissipative structures that all societies build to compensate entropy (disorder). In sum, the uniqueness of social systems in the evolutionary principle lies in how they process and transmit information not by means of biological heredity, but by means of language and symbolic codes. Culture is thus the designer of metabolism fund elements and the combinations of flows of energy and materials that make them function and reproduce. However, culture also produces and reproduces the flows of information that order and give structure to energy and materials flows. This does not mean, however, that there are no entropic costs – whether material, social or regulatory – of the physical consequences of the transmission of information.

While biological systems have a limited capacity for processing energy – mainly endosomatically – due to availability in the environment and genetic load, human societies exhibit a less-constrained dissipative capacity that is only limited by the environment. Human beings can thus dissipate energy by means of artifacts or tools, i.e., through knowledge and technology, and they can do it faster and with greater mobility than any other species. Societies adapt to the environment by changing their structures and frontiers by means of association, integration, or conquest of other societies, something biological organisms cannot do. In other words, contrary to biological systems with well-defined boundaries, human societies can organize and reorganize, building a capacity to avoid or overcome local limitations from the environment. That explains why some societies maintain exosomatic consumption levels beyond the means provided by their local environments without entering into a steady state. The exosomatic consumption of energy is a specifically human trait. Because no genetic load regulates such exosomatic consumption, it becomes codified by culture, which involves a faster but less predictable evolutionary rate.

Human societies give priority to two basic tasks: on the one hand, producing goods and services and distributing them among individual society members, and on the other hand, reproducing the conditions that make production possible in order to gain stability over time. In thermodynamic terms, this implies building dissipative structures and exchanging energy, materials, and information with the environment so that these structures may function. A large number of social relations are geared toward organizing and maintaining this exchange of energy, materials, and information.

1.3 A SOCIOECOLOGICAL VIEW OF SOCIETY: SOCIAL METABOLISM

The organization of this stable exchange of energy, materials, and information has been called *social metabolism*. In other words, social metabolism pertains to the flow of energy, materials, and information that are exchanged by a human society with its environment for forming, maintaining, and reconstructing the dissipative structures allowing it to keep as far away as possible from the state of equilibrium (González de Molina & Toledo, 2014). Open systems such as human societies have managed to create order by ensuring an uninterrupted flow of energy from their environment, transferring the resulting entropy back to their surroundings. From a thermodynamic perspective, the functioning and physical dynamics of societies can be understood on the basis of this metabolic simile: any change in a system's total entropy is the sum of external entropy production and internal entropy production owing to the irreversibility of the processes that occur within.

$$\Delta S_t = S_{in} + S_{out} \qquad \text{(eq. 1)}$$

where
 ΔS_t is the increase in total entropy
 S_{in} is the internal entropy and
 S_{out} is the external entropy

To put it another way, order is generated within a society at the expense of increasing total entropy through the consumption of energy, materials, and information by its dissipative structures or fund elements. This level of order will remain constant or will increase if sufficient quantities of energy and materials or information are added, creating new dissipative structures. This will in turn increase total entropy and, paradoxically, will reduce order or make it even more costly. Complex adaptive systems have resolved this dilemma by capturing the required flows of energy, materials, and information from their surroundings to maintain and increase their level of negentropy, transferring the entropy generated to their surroundings. In other words, the total entropy of the system tends to increase, reducing at the same time internal entropy, if external entropy increases. To put it another way:

$$\Downarrow S_{in} = \Uparrow S_{out} \qquad \text{(eq. 2)}$$

Entropy is reduced by extracting energy and materials from one's own environment (domestic extraction) or by importing from another environment. The greater the flow of energy and materials extracted from its own territory or imported from others (or both at the same time), the more complex an order a society will create, increasing its metabolic profile. For example, physical structures consume resources – both for their building as for their functioning – and have been built for providing health, education, security, food, clothing, housing, transportation, etc. The magnitude of dissipative structures determines the amount of energy and materials consumed by a society, that is, its metabolic profile. Each and every structure needs inputs of a determined amount of energy, materials, and information, and evacuation to the environment of the generated wastes. Differences in countries' installed dissipative structures explain their differences in terms of resource consumption (Ramos-Martin, 2012, 73), and therefore, the differences in the sizes of their metabolisms. These differences in the capability of generating order also indicate the differences in the levels of economic and social well-being.

Consequently, a society's level of entropy is always a function of the relationship between internal and external entropy and, therefore, it is a function of the natural asymmetrical relationship established between a society and its environment, or between one society and another. This asymmetric relationship can even be transferred, as we will see below, to the relationship between different groups or social classes. This is not to say that this relationship is proportional or that an increase in one will always give rise to an increase in the other. To understand this, there is a useful distinction between "high-entropy" and "low-entropy" dissipative structures. A society that requires low amounts of energy and materials to maintain its fund elements reduces its internal entropy, generating in turn low entropy in its environment; in other words, low levels of domestic extractions and/or imports. In this case, the society would produce low total entropy. In contrast, another society might need large amounts of energy and materials from its environment and, if these are not sufficient, it might need to import energy and materials on a large scale in order to reduce its internal entropy. In this case, such a society would generate a much higher level of total entropy. This asymmetrical relationship between society and the environment also translates into differentials of complexity between environment and system, whereby the system is always much less complex than the environment. This forms the basis of the strategy of "biomimicry" (Benyus, 1997) developed intentionally by humans and other high-order species in the extraction of information from the environment, and unintentionally by other living organisms. In fact, biomimicry is perhaps the most determining basic principle underlying Agroecology. In this sense, social metabolism provides Agroecology with a powerful tool for analysis and a theoretical support capable of grounding the hybrid nature – between culture, communication, and the material world – of any agroecosystem, whose dynamics are explained by the interaction of rural societies with their environment (see below).

We can translate this to the accountancy language of the social metabolism methodology (see Schandl et al., 2002): the total level of entropy of a society is assessed using the proxy of the total energy it dissipates during a given year. Dividing the members of the equation makes it easier to make comparisons among societies and provides their metabolic profile. Hence, the level of entropy will be equal to annual domestic energy consumption (DEC year^{-1}), and the metabolic profile will be equal to annual domestic energy consumption per capita per year (DEC population number^{-1} year^{-1}). Accordingly, Prigogine's equation as formulated in equation 1 would take the form:

$$DEC = DEI - Ex, \qquad \text{(eq. 3)}$$

becoming
$$DEI = DE + Im$$

then
$$DEC = DE + Im - Ex$$

where DEC = domestic energy consumption, DEI = direct energy input, Ex = exports, Im = imports, and DE = domestic extraction.

On one side, societies with funds needing low dissipation of energy and low consumption of materials for their maintenance are low-entropy societies that generate low levels of entropy in their environments, i.e., low levels of domestic extraction or imports. On the other side, societies can only sustain high levels of total entropy if large amounts of energy and materials are appropriated from their domestic environment, and if these were locally insufficient, from large imports of energy and materials. This asymmetric relation between society and the environment is also equivalent in differentials of complexity between the environment and the system.

Social practice and social relations are not explainable by the analysis of energy and matter alone, but are not explainable without the analysis of such flows. Reciprocally, social relations represented by flows of information order and condition material exchanges with nature. In other words, the material relations with the natural world that connect human beings with their biophysical environment are a *dimension* of social relations, and as such do not account for its entirety. The specific realm of socioecological relations is the space of intersection between the social sphere – whose structures and functioning rules are self-referential – and the natural sphere, also with its own evolutionary dynamic. Hence, we theorize about the material structure, functioning, and dynamics of human societies on the grounds of a thermodynamic understanding of human societies as biological systems, which they also are. Such a thermodynamic conception lies in the key role played by entropy, both in relations with the biophysical environment and within societies themselves.

1.4 SOCIAL ENTROPY

Some academics have suggested that social dynamics could also be understood in line with the second principle of thermodynamics. Among them is Kenneth Bailey (1990, 1997a, 1997b, 2006a, 2006b), who elaborated a Theory of Social Entropy, attempting to measure it through an indicator of a society's internal state: the level of disorder as a temporal variable. Bailey's approach is also based on the statistical interpretation of entropy of Boltzmann (1964). Statistical entropy may refer to the degree of disorder in social actors' interactions and level of communication (Swanson et al., 1997, 61). Following Adams (1975), a broad concept of energy is applicable here, such as *capacity for performing work* or its physical equivalent *potential energy*, i.e., the capacity for modifying things. Asides from the physical flows already considered (physical entropy), this faculty is possessed by flows of information capable of creating dissipative structures that revert social entropy, a synonym of disorder (Boulding, 1978).

Indeed, for Niklas Luhmann (1986, 1995) human beings (psychic systems) do not belong to social systems but to nature as biological entities, an animal species with special characteristics. Social systems are exclusively made up of communication and function by generating knowledge, i.e., symbols. Psychic systems do not communicate directly among themselves – because their nervous systems cannot interact directly – but through a social system and in doing that, they reproduce that social system. All communicative acts are inherently social and vice versa: there cannot be any communication outside the social system. The components of the social system are precisely the communicative acts, given that systems are built from communication, which starts: "by an alteration of the acoustic state of the atmosphere" (language) (Echevarría, 1998, 143). In that context, entropy is defined as the uncertainty of communication and is in reality an inverse measurement of information. For its part, information, according to Shannon, is the measure of the reduction in statistical entropy, i.e., of disorder (Mavrofides et al., 2011, 360).

Morin (1977) said that under certain circumstances, interactions become interrelations – as associations, unions, combinations, etc. – and generate forms of social organization. In contrast, egotistic, *free rider* behaviors are the opposite of cooperation and tend to generate conflictive frictions. Individuals, social groups, and nation states can adopt a competitive behavior. Social disorder must not be identified with anarchy, but with the total absence of cooperation, which makes it very difficult to organize the activities for social and ecological reproduction, i.e., to maintain the flow of social metabolism. It is not possible to maintain in a stable way the metabolic activity (which ensures the required distance with respect to the thermodynamic equilibrium) without a certain degree of social cooperation.

Continuing with the analogy and adopting an isomorphic perspective regarding thermodynamic laws, social relations can be understood as *frictions* (a term used in Tribology) between social actors, be the latter individuals or institutions (Santa Marín & Toro Betancur, 2015). These frictions are, therefore, uncoordinated and uncooperative interactions that have an effect on social organization. The impossibility of cooperation is what makes societies' survival unviable, or in analogy with thermal death, it leads to *social death*. Therefore, societies in thermodynamic equilibrium are societies in which living in coexistence is impossible. Social disorder would result from social friction motivated by divergent interests, or by competition for scarce resources, i.e., from social conflicts. In that sense, the asymmetries in the allotment of goods and services are, and have always been, powerful stimulators of conflictive frictions. The social energy that becomes degraded and cannot be reinvested in social work can translate into social protests, violent clashes, criminality, bureaucracy, and above all indicators, into exploitation and lack of cooperation. The permanent character of social frictions or conflicts distinguishes this conception from a radical structuralist and functionalist perspective.

Our proposal is far from Bailey's approach (1990), in which social entropy is the product of the dysfunction caused by assigning *macrosocial* mutable factors (population, territory, information, life standard, technology, and organization) to *microsocial* immutable characteristics of individuals (skin pigmentation, sex, and age). While Bailey's characterization of social entropy is far removed from functionalism, it shares its forbiddance of individuals' and groups' capacity to perform, i.e., it omits the capacity of individuals and social groups to change (neopoiesis). Put another way, Bailey's conception places human beings in a condition of alienation. In any case, a more congruent conception with socioecological reality must arise from recognizing that human actions, either individual or collective, are capable of increasing the social system's total entropy, or of decreasing it, producing order. *Entropy and negentropy are possible results of human actions and practices.* Negentropic conflicts generate highly efficient forms of cooperative coordination that drive dissipative institutional structures, replacing entropy by information. Conversely, entropic conflicts destroy forms of cooperative coordination, increasing social entropy. As shown by Robert Axelrod, cooperative coordination is the most efficient and adaptive way of crowning social interaction (Axelrod, 2004). A large part of socioenvironmental conflicts such as the struggles of peasant and indigenous communities against mining extractivism are examples of negentropic conflicts; the destruction of common resources and the privatization of land are examples of entropic results.

In short, strictly speaking, no social entropy or a second principle of thermodynamics governs "social energies". We consider "social entropy" as an analogy that highlights the key role of information flows in the fight against entropy. Information is used by human beings to build and operate the dissipative structures that keep society away from thermodynamic equilibrium. In that sense, from a physical perspective, certain types of social relationships constitute information flows that prevent individualistic or egoistic behavior from putting an end to society. For example, social relations favoring cooperation are much more efficient than those favoring non-cooperation. In terms of Tribology, cooperation decreases the wear caused by increased frictions (inherent to complexity) by means of designing such frictions (institutional relations) and of their lubrication (motivation) in stimuli and penalizations. Friction is the result of a conflict, or better, friction *is* conflict: a relation of competition and confrontation between two individuals or groups. Frictions (conflicts) can be regulated cooperatively or be deregulated non-cooperatively. The underlying wear in non-cooperative deregulation is much more severe and the motivation much weaker than in cooperative regulation. Observe the difference between the ordered evacuation of a crowd from inside a stadium guided by rules, signals, and counting, with accessible exit areas, and the same crowd evacuating the stadium by means of chaotic movements against one another. The amount of frictions (physical contacts) and the amount of wear decrease in an ordered evacuation compared to a chaotic evacuation.

An example of misunderstanding of how incoordination leads to high levels of physical entropy can be found in the so-called *tragedy of the commons* (Hardin, 1968), in which the responsibility of grassland is assigned to the community owning it. Certainly, an aggregated set of individual, non-coordinated behaviors leads to an unsustainable level of exploitation of any resource; but as shown by E. Ostrom (2015a), what leads to overexploitation is the lack of communal management, not vice versa. Individual, non-coordinated action is an example of maximum friction generating an increase of physical entropy, but also the increase of social entropy in the long term due to growing inequity and scarcity of resources. The granting of property rights (a market alternative proposed by Hardin) or the centralized management of an external, coercive regulator (stratification alternative) are possible answers to the problem of individual incoordination. Yet, as shown by Ostrom (2015a), they carry their own dose of entropy from incentivizing competition and inequity. Communal, cooperative management of resources is the form of coordination that generates least social, political, and physical entropy because it minimizes social frictions, and with that, disincentives non-cooperative behaviors.

Put another way, the information within systems has the function of establishing coordination and cooperation subsystems that reduce frictions, hence also entropy.

A society's "thermal death" would be equivalent to the total absence of cooperative behaviors that would make the functioning of the metabolism with the environment impossible, leading to thermal equilibrium. Information flows therefore have the mission of ensuring the necessary coordination between individuals so that metabolic activity may take place.

1.5   INSTITUTIONS AND SOCIAL INEQUITY

As seen above, changes in a system's entropy are always asymmetrical, hence unequal, relative to both terms of the equation: internal and external entropy levels. The environment and its resources, either in domestic or in *other* society's territory, pay the costs of this asymmetry. Asymmetry is thus at the core of each dissipative process because these processes operate following two contrary directions: producing work (order) and generating unusable heat (disorder) (Hacyan, 2004). Inequality is thus the bucketing of order in one direction, and of disorder in the opposite direction. This dichotomy also powerfully stimulates the interactions between individuals and groups in their quest for more energy and materials to maintain order and decrease disorder. In that context, a major part of social relations is aimed at exchanging energy, materials, information, and wastes.

Therefore, inequality between social groups is a socially established mechanism of transference of entropy, which may generate more entropy if not counterweighed by more energy and materials from the environment or by socially built negentropic structures. It also means that rising social complexity is often the result of social inequity or, said differently, as inequality increases – a process apparently peaking at present – more energy and materials have been consumed, increasing complexity. Why are capitalism and its industrial metabolic regime based on increased inequality? Because they require transferring the high entropy they generate to their social or territorial surroundings.

Consequently, asymmetry is applicable to relations between groups or classes within a society and has direct consequences on their environment. For instance, a social group can push toward the overexploitation of one or more resources if it accumulates or consumes a growing proportion of the energy and materials available to a society within its territory. In other words, the creation of internal order by a human group can have consequences on the society's environment as a whole. An example makes this fact more graphical: in feudal or tributary societies based on organic metabolism, rent increase forced peasants to offer a portion of their crop, or other natural resource, to the detriment of the amount needed for self-consumption, and could push them to clear new plots, fish or hunt more individuals, and extract or gather a higher volume of products.

Following the thermodynamic analogy, we could reformulate Prigogine's equation to apply it to social systems: the change in social entropy of a given human social group is directly related with (physical) entropy and its distribution through society. For example, units of social organization with proper coherence and identity (e.g., social classes) can increase their internal order or well-being by transferring their (physical) entropy to other social classes or to society as a whole. The appropriation by a social group of the surplus (a determined amount of energy or materials) is but a way of obtaining order at the expense of other social groups that will experience an increase in their level of entropy. It is a form of *socio-thermodynamic exploitation*. Such has been the usual behavior of societies, according to Flannery and Marcus (2012), at least since the past 2,500 years, based on competitive social relations and the institutionalization of social inequity. Since then, conflicts and inequality seem to have amplified throughout history until the present. The same can be said about the asymmetric relation between dominant and dominated kingdoms or states, which becomes evident in industrial metabolism, countries, and social classes showing marked differences in levels of social inequity. On the contrary, the predominance of cooperative social relations and institutions favor equity aids to reduce "social entropy". An obvious relationship can be established between social and physical entropy. Increasing physical entropy has been one of the most recurring ways of compensating for rising social entropy, as we will see further down.

In view of the asymmetrical relationship, human societies have built dissipative structures of a social nature (see below) based on cooperation and equity, without which social life and evolutionary success itself would be impossible, given that maximizing asymmetry would lead to disorder or thermodynamic equilibrium. Nevertheless, *free-rider* behaviors are common among human groups, who, to maximize their order, increase entropy throughout society, those deprived of enough dissipative structures being the most damaged. This behavior is evident both in the struggle for resources (energy, materials, and information) as in the fight to avoid the effects of entropic disorder (e.g., pollution). This is when conflicts arise from power relations (between individuals or groups) and control relations (of individuals or groups over the flows of energy and materials; Adams, 1975).

1.6   POLITICS AND ENTROPY

All species have developed phenotypic (endogenous) plasticity mechanisms with which to respond adaptively to changes in the environment. The human species has gone much further and has managed, using exosomatic or technological mechanisms, to modify its own environment and somewhat adapt it to its own interests to such an extent that the environment has become vulnerable. One of these technological mechanisms has consisted in creating coordinated and intentional interaction systems, that is, institutions. Paraphrasing Ian Wright (2005), we could say that a society consists of vast numbers of people constantly interacting and huge numbers of degrees of individual freedom. Its behavior could be considered as similar to that exhibited by randomizing machines that maximize entropy, subject to constraints. Entropy is here understood as a number that measures the randomness of a distribution. In this sense, the higher the entropy, the more random the distribution. The individual behavior randomizes the distribution of social energy in the system and increases its (social) entropy. However, society builds social institutions (dissipative structures made of information fluxes) to prevent entropy maximization or to reduce it. As Wright argues, "at a micro level the system scrambles and randomizes. Basically, anything can happen. But at the macro level there are global constraints that are always observed. So there's an interaction between forces that randomise, and forces that order … There's some kind of obstacle in the system that acts to reduce the randomness a little bit". This is what happens, for example, in completely unregulated markets: they tend to maximize the system's entropy. According to Wright (2017), this fact could explain the inequality: "So the anarchy of the market is the primary and essential cause of economic inequality ... Since people are free to trade then entropy increases and the distribution of money becomes unequal". Regulations thus add constraints to the system, reducing entropy, that is to say, reducing random (unequal) distribution.

Institutions can thus be defined as formal (explicit) or informal (implicit) rules (routines, procedure, codes, shared beliefs, practices) aimed at regulating the entropy that arises from the coordination of social interactions between individuals (Schotter, 1981) and between them and their natural environment. This understanding has a twofold explanatory and regulatory dimension: entropy indicates both the evolutionary origin of the institutions and the teleological function they fulfill. A level of entropy, disorder, and loss of work in the transformation process can be found in all types of interactions there, but in the case of living systems intentionality, reflexive or not, adds to the interaction (Dennett, 1996), increasing the complexity. In the case of human social systems, where mechanisms of natural cultural selection operate, regulating entropy becomes more sophisticated and is expressed in flexible and contingent rules through institutions. Of all institutions, we are especially interested in political institutions. Our interest lies in the fact that they represent the highest degree of complexity in their regulation of social interactions, on a scale of interactive density that is equally high.

In our proposal of agroecological theory of politics, entropy would explain the causes, functions, and mechanisms that create and give meaning to forms of power (regulation) from micro to macro (State) levels of political power. Family, community, State, are examples of negentropic structures and so is any efficient form of social regulation. As a consequence, social institutions in the broadest sense – understood as any stable social practice or relation subjected to rules, although these may be informal – and also in the narrowest sense of state public institutions, must be seen as socioecological relations tasked with regulating both social and physical entropies. In other terms, political power manages entropy by means of generating dissipative structures in the physical, political, and social realms.

1.7 POLITICAL INSTITUTIONS: THE TRADE-OFF BETWEEN SOCIAL AND PHYSICAL ENTROPY

An entropic theory of regulatory institutions claims that there is an isomorphism between the three dimensions of entropy (physical, political, and social; Figure 1.1), such that more social entropy (inequity) is corresponded by more physical entropy. Therefore, the function of political regulators is to synchronize social metabolism at its two extremes (biosphere and society) in the knowledge that the regulatory function itself implies an inherent entropic cost, hence regulating the entropy generated by regulation itself. This confers a high degree of complexity and self-reflexivity to political institutions that cannot be substituted by simple self-management mechanisms.

Indeed, the effects generated by social and physical entropy, i.e., the type of disorder they trigger, are different. Physical entropy expresses itself as ecological

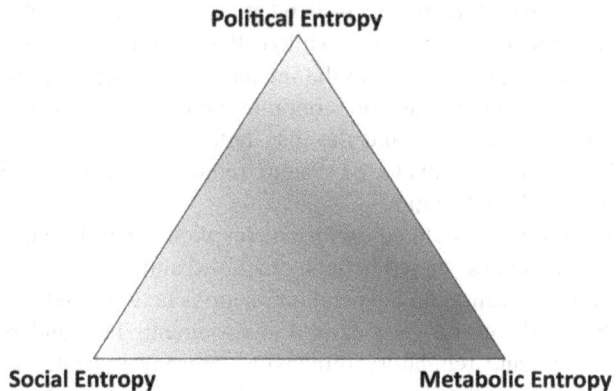

FIGURE 1.1   The Triangle of Entropies.

problems and its effects are of a physical nature. Social entropy expresses itself as social conflict and destructuration (confrontations, inequity, and absence of cooperation, criminality, and poverty). However, there is an evident trade-off between social and physical entropies that, from our standpoint, is useful for explaining the evolutionary dynamics of social systems. Social entropy translates (or rather *transduces*, a neurology term for communication mechanisms that convert environmental signals into neurophysiological states) the level of physical entropy, and in turn, social entropy produces a certain level of physical entropy. Thus, there is a bidirectional correlation between both types of entropy, which can be formalized by once again using Prigogine's equation 1 expressed above:

$$SESt = (SSin + PhSin) + (SSout + PhSout) \qquad \text{(eq. 4)}$$

where *SESt* = socioecological entropy, *SSin* = internal social entropy, *PhSin* = internal physical entropy, *SSout* = external social entropy, and *PhSout* = external physical entropy.

For example, inequity produces situations that tend to increase social entropy, e.g., bringing a society or a social group closer to disorder or chaos, generating poverty or scarcity or even starvation. Usually, societies compensate such increment by importing a certain amount of energy and materials from the environment to generate order. For example, rising exosomatic consumption over the past two centuries has been the system's answer to growing inequalities threatening to raise social entropy to unsustainable levels. In this interpretation of social entropy, exosomatic consumption turns into an instrument compensating the maintenance of an unfair social system through the construction and installation of new, more costly, dissipative structures, reducing internal entropy and in parallel raising external entropy, i.e., transferring entropy to the environment.

This bidirectional relation is mediated by regulatory or institutional mechanisms, which are also subject to the principle of entropy (political or institutional entropy). Inefficient social allotment of resources generates a reaction of tension (conflict) and the associated information generates negentropic structures (institutions), which are but filters, sensors, or institutional programmers responsible for detecting and compensating social entropy by means of physical entropy, or vice versa. Therefore, political power, the regulator and producer of information and coordination, has a negentropic function. The paradox lies in the fact that this negentropic function generates its own entropy. The positive equilibrium between negentropy and entropy marks the limits of validity and success of a determined form of political power. Political power reduces entropy by means of fostering the coordination between the different stakeholders (individuals and institutions) involved in social metabolism. Cooperation is in itself the less-entropic form of coordination (Axelrod, 2004). The democratic State and society represent a form of cooperative coordination that can occur in societies with high demographic and technological complexities.

Until the first symptoms of the ecological crisis appeared, capitalist societies reduced their levels of both types of entropy – social and political – by transferring it to the biophysical environment, hence increasing physical entropy:

$$\blacktriangledown PS + \blacktriangledown SS \rightarrow \Delta PhS \qquad \text{(eq. 5)}$$

where: $\blacktriangledown PS$ = decrease in political entropy, $\blacktriangledown SS$ = decrease in social entropy, and ΔPhS = increase in physical entropy.

However, the environment of all societies, i.e., the natural part formed by the biosphere, atmosphere, hydrosphere, etc., is a closed system receiving energy only from the outside, not an open system exchanging energy and material with its surroundings. Thus, the possibilities of compensating for social and political entropy through the establishment of more and larger dissipative structures of energy and materials, i.e., by elevating physical entropy, have obvious environmental limits. In fact, the current economic–financial crisis is a clear example of the unfeasibility of this compensation model, because the crisis is caused by the difficulties of maintaining economic growth in a context of increasingly limited natural resources (see Chapter 3). The impossibility of further increasing physical entropy is responsible for the rising levels of social and political entropy, even in developed countries. Therefore, political power has to manage political entropy (distribution of power and status), social entropy (distribution of resources), and physical entropy (exchange between society and nature), and at the same time, the interactions between all three entropies. From the latter perspective, the liberal myth of the radical separation between political society and civil society is further unjustifiable.

In sum, the function of political institutions is to control and minimize physical and social entropies by means of flows of information, but also by means of managing its own internal entropy (transaction costs, bureaucracy, political oligarchies, centralized decisions, war, etc.). The entropic propensity of political institutions (Niskanen theorem) is an unavoidable price institutions must pay due to the nature of negentropic institutions in the physical and social realms. In that sense, it is important to distinguish between the operative entropy of political regulators (highly entropic institutional designs such as dominant valuation languages in conventional economics, neoclassical macroeconomic indicators such as gross domestic product (GDP) or the use of market prices as a unique indicator of value) and the political regulator's internal, functional entropy that translates into centralized decisions, high costs of transactions, bureaucracy, etc. Political institutions reduce social and physical entropy by increasing their internal entropy. That makes political proposals characterized by *minimal statism* or deregulation dangerous (entropic). These proposals may be somewhat popular or intuitively successful because they alleviate us from the State's strong entropic propensity and that of regulatory institutions, but the risk is greater still: that of increasing social and physical entropy.

1.8   CONFLICT, PROTEST, AND METABOLIC CHANGE

Some social relations or interactions create organizational forms that become sources of conflicting frictions, and thus of potential disorder, for example: unequal allotment of goods, services, and wastes; decoupling between population size and resource availability; reproductive failures; or the patriarchal system of discrimination by the sex–gender subsystem. In this sense, different forms of social inequity are the main source of conflictive friction and, therefore, of "social entropy". Conflicts originate in this context of unequal distribution of resources. Its entropic or negentropic resolution will depend on the cooperative, coercive or competitive orientation finally adopted. Social inequality is expressed by the unequal assignation of goods and services among social groups or their territories, thus creating hierarchic societies. A physical interpretation of this social disequilibrium is the unequal distribution of flows of energy and materials, and of recycling of wastes – i.e., the ecosystems' absorption service. In the absence of equalization forces, the random (entropic) tendency as a universal biophysical law tends toward the concentration of resources in a tiny minority of species or social classes, both in biological ecosystems and in social systems. These equalizing forces are "natural enemies" in ecosystems and institutions in social systems. If, for various reasons, natural enemies disappear or decline, and institutions called to regulate social entropy are inhibited or specialize in fostering social asymmetries, inequality expands (Scheffer et al., 2017). Social death or society's thermal equilibrium would be the result of maximum friction caused by an extreme inequality in the distribution of metabolic flows that guarantee the distance from thermal death. Just as the most entropic metabolic regimes are those that work with a high degree of energy dissipation, the most entropic social systems are those that lack cooperative institutions concentrating resources in a small portion of society. Social entropy (inequality) is thus the result of a lack of cooperation that prevents or restricts access to services provided by dissipative structures created by society.

Certainly, the unequal access and distribution of resources in a broad sense, including material and immaterial resources, has historically been a permanent source of conflicts and social protest. However, they have also been, paradoxically, a powerful driving force of the historical evolution of societies and their metabolic configuration. Social protest emerging from conflicts is one more factor thrusting metabolic change and, according to the historical circumstances, can even become the most decisive of all. Consequently, they must be taken into account when studying the evolutionary dynamics of social metabolism and the socioecological relations between different human groups. The same must be said about the dynamics of agroecosystems.

Hence, social protests have a contradictory impact on their environment in the form of entropy or negentropy: they can produce order or disorder, increase or decrease social and physical entropy. One example is the positive conservation effect of the communal defense of forests pursued for a long time by indigenous communities, removing forests from markets and avoiding their clearing; in contrast, the struggle of peasant laborers in Spain and Italy during the mid-twentieth century, within an uncontested framework of capitalistic competition, caused an increase in labor costs due to a rise in wages, which indirectly and unintentionally favored the mechanization of most agrarian tasks. This mechanization implied the use, by agriculture, of massive amounts of fossil fuel. There was, despite the environmental effects, a tangible improvement in peasants' poor living conditions.

Collective action is a basic component of the autopoietic, and even neopoietic capabilities of social systems. It is often guided by common objectives of the individuals participating in the conflict. Thus, from the standpoint of intentionality, collective action can promote the construction of dissipative structures to decrease internal entropy (disorder), also reducing or transferring the external entropy to the physical environment. The crucial factor is the character of dissipative structures (high or low entropy) built by a self-organization process promoted by collective action. In that sense, and differently from conflict, social protest as a collective action cannot be understood as a part of disorder itself, but as a generator of negentropy. Consequently, a protest rising from social conflict and guided by an agenda to change the dominant metabolic regime may give rise from its initial stages to the gestation of dissipative – negentropic – structures that decrease internal disorder as well as, simultaneously, the consumption of energy and materials, so that the transference of entropy to the environment – i.e., the external entropy – is minimized. Disorder through social protest (high-quality, low-entropy information) can generate a new, self-organized and coherent emergent order.

Conflicts can balance or further unbalance a social group's internal and external entropies. Usually, protest arising from an environmental conflict – especially ecologist protests – help internalizing environmental costs and, while they do not succeed in changing social metabolism instantaneously, they improve the negative effects on the environment and open the way toward metabolic transformation. They thus function as reducers of the system's internal entropy, i.e., they reduce the amount of entropic flow that is transferred to the physical–biological environment. In that sense, environmental protest can generate social actions promoting a change in the structure and composition of a more sustainable social metabolism, but they can also promote the appropriation and use of more energy and materials, raising external entropy to the level of total entropy, as is the case in most modern armed conflicts between nation states or their coalitions. However, the separation between social (class) and environmental (species) conflict, a typical dissociation of the industrial metabolic regime, has obvious risks: social conflict may end up increasing metabolic entropy (environmental degradation), while environmental conflict may end up increasing social entropy (inequality), both entropies growing in a fatal loop of negative synergies. In this perverse way, the dissociation between environmental and social conflict can reverse the negentropic nature of these conflicts into an objectively entropic outcome. For example, conflicts over wage improvements after the Second

World War were compensated by substituting human labor with capital (machines), increasing the consumption of energy and materials. Today, attempts to reduce greenhouse gas emissions by increasing fuel prices hurt workers who live on the outskirts of cities with restricted access to public transport or provide transport services in small enterprises. When protest only considers environmental damage and not social damage or vice versa, the risk of this happening is very high.

1.9 POLITICS IN AGROECOSYSTEMS

As we have seen, the theoretical framework developed so far could be applied to socioecological relationships, whatever their scale or territorial scope. It can therefore be applied to agroecosystems and to food systems as a whole.

In doing so, agrarian social metabolism (ASM) or *agrarian metabolism* (AM) can be described as the exchange of energy, materials, and information that agroecosystems perform with their socioecological environment. The purpose of metabolic activity is that of appropriating biomass to satisfy human species' endosomatic consumption directly or indirectly through livestock while providing basic ecosystem services. AM has also tried to satisfy societies' exosomatic demands (raw materials and energy) with an organic metabolism and continues to do so, to a lesser extent, in industrial societies. To accomplish this, society colonizes or seizes a part of the available land. Within this territory, it establishes varying degrees of intervention or interference in the ecosystems' structure, functioning and dynamics, giving rise to different types of agroecosystems. In other words, AM refers to the appropriation of biomass by members of a society by managing the agroecosystems present on the land (Guzmán Casado & González de Molina, 2017).

According to the thermodynamic approach that we have been following, agroecosystems can also be considered as complex adaptive systems that dissipate energy to counteract the law of entropy (Prigogine, 1978; Jørgensen & Fath, 2004). To do this, they exchange energy, materials and information with their environment (Fath et al., 2004; Ulanowicz, 2004; Jørgensen et al., 2007; Swannack & Grant, 2008). Compared to ecosystems, which still retain their capacity to self-sustain, self-repair, and self-reproduce, agroecosystems are unstable, requiring external energy, materials, and information (Toledo, 1993; Gliessman, 1998). The flows are exchanged through work or manipulations that aim at ensuring the production of biomass and its reiteration over successive cycles of cultivation or breeding, interfering in the carbon, nutrients and hydrological cycles and in the mechanisms of biotic regulation. In traditionally managed agroecosystems, this input of additional energy and materials comes from biological sources: human work and animal labor. Dependence on the land is maintained in a strict sense. In industrially managed agroecosystems, additional energy and materials also come from the direct and indirect use of fossil fuels as well as metallic and non-metallic minerals. In short, they are part of society's general metabolism, specifically dedicated to the appropriation of photosynthesis products.

From a metabolic point of view, the reproductive dynamic of agroecosystems is peculiar. Their sustainability, as artificialized ecosystems, also depends on their level of biodiversity, the maintenance of a fertile soil, etc. This means that part of the generated biomass must be recirculated to meet both productive and basic reproductive functions of the agroecosystem itself: seeds, animal labor, organic matter in the soil, functional biodiversity, etc. The thermodynamic rationale underlying this characteristic was developed by Mae-Wan Ho and Robert Ulanowicz (2005), and later by Ho (2013), when they related sustainability to dissipative low-entropy structures. Ecosystems, as dissipative structures that can consume large amounts of energy or the reverse, can be structured in such a way that their entropy is low. This characteristic of ecosystems also works at different scales for agroecosystems and even for AM as a whole. Like ecosystems, agroecosystems constitute an arrangement of biotic and abiotic components in which living systems predominate and respond to what has been called "thermodynamics of organized complexity" (Ho & Ulanowicz, 2005, 41, 45). This means that, going beyond the point raised by Prigogine (1962), an agroecosystem can be "far from thermodynamic equilibrium on account of the enormous amount of stored, coherent energy mobilized within the system, but also that this macroscopically non-equilibrium regime is made up of a nested dynamic structure that allows both equilibrium and non-equilibrium approximations to be simultaneously satisfied at different levels". In this sense, the really decisive aspect of ecosystems is not only the flows of energy and materials that keep them away from thermodynamic equilibrium, but also their capacity to capture and store the energy that circulates inside them and transfer it to its different components (Ho & Ulanowicz, 2005, 41–45). This depends on the quality and quantity of circuits or internal loops through which the energy flows circulate as well as whether they are able to compensate for the entropy generated somewhere in the ecosystem by the negative entropy generated in another system within a given period of time. As Bulatkin (2012, 732) argues, "the agroecosystem as a natural-anthropogenic system has its own biogeocenotic and biogeochemical mechanisms and self-regulation structures, which should be used to reduce anthropogenic energy costs". That is, it contains cycles that, according to Ulanowicz (1983), have a "thermodynamic sense": "Cycles enable the activities to be coupled, or linked together, so that those yielding energy can transfer the energy directly to those requiring energy, and the direction can be reversed when the need arises. These symmetrical, reciprocal relationships are most important for sustaining the system" (Ho & Ulanowicz, 2005, 43).

For example, in organic or agrarian metabolic regimes (González de Molina & Toledo, 2011, 2014), agroecosystems used to function in an integrated manner. Biogeochemical cycles clearly went beyond the cultivated lands and extended over large parts of the territory. The increase in entropy that occurred in the most intensively cultivated areas (irrigation or hedges, in the case of the Mediterranean) was usually compensated by the import of nitrogen through livestock (manure) from other areas of low entropy such as forest areas. The result was a metabolic regime that was also of low entropy. Spatial heterogeneity and agrosilvopastoral integration was the key for articulating the different circuits that captured, stored and transferred energy.[1]

---

[1] As pointed out by Sieferle (2001, 20), different land uses were linked to different types of energy. Cultivated lands were associated with the production of metabolic energy to provide human food; the pasture land that fed farm animals was associated with mechanical energy and forests with the thermal energy that provided the fuel needed for cooking, heating and manufacturing.

This explains why, when different agroecosystem components are adequately articulated, it is possible to substantially reduce incurred land costs whenever biomass is produced, and thus generate the largest amount of biomass at a minimal land cost (Guzmán & González de Molina, 2009; Guzmán et al., 2011). In this sense, net primary production is found to correlate positively with the functional integration of different land uses in terms of territorial efficiency. The bigger the amount of energy captured and stored in the internal cycles of agroecosystems, the smaller the amount of energy that will have to be imported from outside (Guzmán Casado & González de Molina, 2017). For this reason, it is often commented (Gliessman, 1998) that the more an agroecosystem resembles natural ecosystems in its organization and functioning, the greater its sustainability.

1.10   FUNDS AND FLOWS IN AGRARIAN METABOLISM

We have defined AM as the exchange of energy, materials, and information between agroecosystems and their socioecological environment. This exchange is composed of flows that go in and out. These flows have a dual function: they maintain and make the dissipative structures or fund elements work. The distinction between flows and funds was borrowed from Nicholas Georgescu-Roegen (1971) and Mario Giampietro and colleagues (Giampietro et al., 2014). According to Georgescu-Roegen, the economy's ultimate goal is not the production and consumption of goods and services, but the reproduction and improvement of the processes necessary for their production and consumption. This different understanding of economic activity's main objective implies that from a biophysical point of view, we need to shift our attention away from energy and material flows and instead focus on fund elements, in this case AM: we must pay attention to whether fund elements are improved or at least reproduced during each productive cycle. In other words, our focus switches from the production and consumption of goods and services to sustainability, and whether both production and consumption can be maintained indefinitely.

Flows include energy and materials that are consumed or dissipated during the metabolic process, such as raw materials or fossil fuels. The rhythm of these flows is controlled by external factors – relating to the accessibility of the environment's resources in which the metabolic activity unfolds – and by internal factors – related to the processing capacity of energy and materials, relying in turn on the technology used and the knowledge to manage it. Fund elements are dissipative structures that use inputs to transform them into goods, services and waste, i.e., into outputs, within a given time scale; they remain constant during the dissipative process (Scheidel & Sorman, 2012). They process energy, materials, and information at a rate determined by their own structure and function. To do so, they need to be periodically renewed or reproduced. This means that part of the inputs must be used in the construction, maintenance and reproduction of the fund elements, limiting, of course, their own processing rhythm (Giampietro et al., 2008). The quantities of energy and materials invested in the maintenance and reproduction of the fund elements cannot be employed for end uses. These types of elements can even be improved over time, when energy and materials are allocated for this purpose.

Land, livestock, the agrarian population who manages the agroecosystems, and technical means of production or technical capital are all examples of fund elements. Depending on the purpose of the analysis, each of these funds could be split into different funds. For example: land could be divided into various fund elements, e.g., biodiversity, ability to replenish nutrients, organic matter content, etc. In this sense, it is relevant to differentiate between fund elements of a biophysical nature and fund elements of a social nature because they are not reproduced in the same way. All of them are closely connected and represent the fullest manifestation of the socio-ecological relationships at the heart of each agroecosystem and at the center of the metabolic exchange. The articulation between the fund elements is fundamental, as we shall see later, to explain metabolic dynamics.

Depending on its biophysical or social nature, each fund element works with different quality flows and different metrics. Fund elements require a quantity of energy in terms of biomass and human work that must be taken into account for each production process. The process of industrialization of agriculture has consisted in substituting the agroecosystems' biogeochemical circuits with working capital that depends on resources outside the agrarian sector, usually via markets. This explains a fundamental difference in the metabolic functioning of traditional and industrialized agroecosystems: the reproduction of fund elements was possible through biomass flows in organic metabolic regimes; but under the industrial metabolic regime, external fossil energy flows are widely reproduced by social funds and can cause environmental deterioration when attempting to reproduce biophysical funds, especially agroecosystem services. For example, trophic chains that support both edaphic life and other organisms within the agroecosystem's biodiversity can generally only be fed with biomass. Deterioration of colonized or appropriated land cannot be compensated using energy and external materials or any other resource than vegetal biomass. In this way, the industrialization of agriculture can be interpreted as the process of replacing dissipative structures of a biophysical nature, that belong to agroecosystems and have been maintained by peasants through integrated management, with man-made dissipative structures or, to put it in economic terms, with means of technical production obtained through markets and, to a lesser degree, from State intervention.

However, in AM not only are energy and materials flows exchanged, but also flows of information. These flows play a fundamental role: they have the capacity to order and organize the physical, biological and social components of agroecosystems. Indeed, agroecosystems process energy and materials to produce biomass thanks to human labor. As we have seen, human work has a characteristic that makes it pivotal: it incorporates information flows. The origin of these flows is not farmers alone but also the household they are part of. Consequently, the main fund element of agroecosystems is the "agrarian population", composed of domestic groups or households that are dedicated to this activity. There are three reasons for this, based on the distinction between flows and funds. First, because the continuation of the human work flow depends on the time investment in other tasks carried out by the entire household. For example: time devoted to care, which are reproductive tasks from the physiological point of view, or to social and educational activities, which from a social perspective would correspond to reproductive activities.

Second, because maintaining agroecosystems in good productive conditions requires performing maintenance tasks that are not usually considered to be part of working hours directly related to agricultural production or are effectively paid. And lastly, because agricultural labor has usually been performed by farmers with the help of the family, so agrarian work is essentially family work. As a result, we not only considered the number of individuals engaged in agricultural work, but also their families who are responsible for "producing" the agricultural workers and who can engage in other paid and unpaid activities to achieve it. In fact, for peasant and small and medium farmers, the family is above all the basis of their economy and the objective of their productive strategies.

Human labor logically requires energy, basically endosomatic energy, to maintain and reproduce itself. In fact, this is the amount of energy that we used to calculate the energy efficiency of each of the successive metabolic arrangements over time. Nevertheless, as human societies have been gaining in complexity, cost of reproduction has also increased to include all exosomatic energy incorporated in that process (or its equivalent in monetary terms). As the metabolic profile of contemporary societies has increased the cultural consumption of energy and materials, so has its monetary cost. Consequently, the concept of agrarian metabolism takes into account not only the number of individuals engaged in agricultural work and the time spent on it, but also their families or households and the paid or unpaid work time that is required to sustain them. The maintenance of a constant flow of human energy needed to manage agroecosystems depends on the reproduction of these agricultural groups. Reproduction costs must be covered by the own farm (self-consumption), income from the sale of agricultural production or from money obtained from the sale of labor power or other gainful activities. The fourth and last fund element considered is technical means of production. Today it could be called "Technical Capital" as referred to by Mario Giampietro et al. (2014). The maintenance of this fund requires investment in energy and materials and, unlike the other funds, its replacement occurs thanks to metabolic processes that take place outside the agricultural sector itself.

We assumed that peasant or farmer decisions were directly influenced by the ability to cover reproduction costs. They depend, in fact, ever more on the monetary remuneration that they receive in exchange for the sale of their products. Therefore, monetary flows constitute a suitable proxy for synthesized information flows. Information flows are defined as follow: flows originating in the agrarian population fund element, in the form of work and incorporated management decisions; and monetary flows stemming from the agroecosystem's social environment and ending up in this information fund in the form of money obtained in exchange for production. Money, expressed in relative prices, has transmitted information that has enabled to largely explain – especially in societies with monetized exchanges – the behavior of social agents, in our case that of farmers and peasants. This does not mean that markets, as they are organized today, have determined farmers' behaviors based on relative prices. Markets were not always the main or only way to exchange goods and services. Therefore, their dynamics only explain productive decisions in contexts of commodified economies, where relative prices are the most relevant indicator or source of information. We are aware of other non-monetary information flows that also have a bearing on farmers' decisions, but momentary flows are essential, especially for market or capitalist societies.

1.11 THE ORGANIZATION AND DYNAMICS OF AGRARIAN METABOLISM

Moreover, the relative prices of inputs and agricultural products determine the monetary income received by farmers, and therefore, they determine the reproduction possibilities of agroecosystems' fund elements. These prices do not result, as classical and neoclassical economic theory claims, from the intersection between supply and demand, nor are they merely an expression of the exchange value of such commodities, as maintained by Marxist theory. They also stem from regulations and norms that generate a favorable environment to maintain a given configuration of social relations and power relations. For example, in capitalist societies, markets operate within an institutional framework designed to reproduce the socioecological conditions necessary to maintain them as a system of domination. The trade-off between metabolic entropy and social entropy in agriculture has relied on this institutional framework, converting industrialized agroecosystems into dissipative structures with high entropy and low sustainability levels.

The market itself and private property constitute two of its major pillars, but they are not the only ones. They form an institutional conglomerate that directs energy and material flows within the agroecosystems and explains the degree of access to their fund elements. This institutional conglomerate has been described as a *food/alimentary regime*. Regime comes from the Latin word "regimen" and refers to the set of rules that govern or regulate an activity or a thing. Such rules, in turn, reflect specific power relations that aspire to becoming permanent by means of these rules, to endure over time, benefiting those in a position of power or domination. When those norms are maintained over time, we can speak of regime: "The international relations literature also uses the term 'regime' to capture the formation of self-governing networks which enable partners to meet shared concerns. International regimes are systems of norms and roles agreed upon by states to govern their behavior in specific political contexts or issue areas. Regimes are formed to provide regulation and order without resort to the over-arching authority of a supranational government [...] The analysis of international regimes has largely concentrated on the coming together of state actors, although the involvement of non-state actors is not entirely neglected" (Stoker, 1998, 23).

This term has been applied to food systems to highlight the stable nature of institutional arrangements and the power relations that sustain them, especially since the late nineteenth century with the consolidation of nation states and the arrival of the First Globalization (Friedmann, 1987; McMichael, 2009). Indeed, Harriet Friedmann defined the *food regime* as a "rule-governed structure of production and consumption of food on a world scale" (Friedmann, 1993, 30–31). It is not necessary for these rules to be completely explicit, the regime is based on increasingly sweeping international agreements and national legislations that enshrine the empire of private property and the market: "The food regime, therefore, was partly about international relations of food, and partly about the world food economy. Regulation of the food regime both underpinned and reflected changing balances of power among states, organized national lobbies, classes – farmers, workers, peasants – and capital. The implicit rules evolved through practical experiences and negotiations among states, ministries, corporations, farm lobbies, consumer lobbies and others, in response to immediate problems of production, distribution and trade" (Friedmann, 1993, 31). As we will see in the chapter that follows, food regulation has been the expression of governments' economic policies, of the interests of agricultural input companies and of the progressive hegemony of large retail chains and agribusiness.

Friedmann's conception of the food regime is open, far removed from a strict expression of dominant interests: it is dynamic and changing, determined not only by relentless power relations between States and corporations, but also by social movements fighting against that imposed order. The objective is to maintain the essence of the regime while adapting to changing scenarios. The food regime therefore reflects power relations, but it also constitutes a stable way of organizing the flows of energy, materials, and information; it plays a key role in the functioning of social metabolism at a national and international level, given that food is a key element in the trade-off between biophysical and social entropy. The concept of food regime thus arose to designate the series of norms regulating the world's food system. As explained above, it can, however, be applied nationally and even locally. The concept is applicable at almost any territorial scale.

1.12   AGROECOLOGICAL TRANSITION AND FOOD REGIME CHANGE

As we will see later, an agroecological change is not possible without a change in the institutional framework. A change of the dominant food regime is the main objective of Political Agroecology. In the social sciences, theories that explain long-term changes in human societies using the concept of transition have become increasingly relevant (Bergh & Bruinsma, 2008; Lachman, 2013). A theoretical proposal that analyzes transition toward sustainability from a metabolic perspective has also been developed, given that the ecological crisis pushes toward a better understanding of historical change, in this case, toward sustainability. According to this approach, socioecological transitions constitute processes of structural change affecting the configuration of energy, material, and information flows that societies exchange with their environment (Fischer-Kowalski & Rotmans, 2009; Fischer-Kowalski, 2011). Following this approach, the food regime transition could be understood as a process of shifting from one institutional framework to a qualitatively different one. Such a transition implies major changes, and not simple readjustments or improvements. It means attaining a new, qualitatively different food regime. In this sense, we agree with Fischer-Kowalski and Haberl (2007, 7) when they state that the socioecological transition is the product of a deliberative change. This claim is supported by the transition to a more sustainable world: it seems logical, but it is far from being inevitable.

We are adopting an instrumental and contingent approach to the food regime, thus lessening its normative load, not only because it is an *ex post facto* created tool for understanding historical processes, but also because socioecological change is a constitutive property of social systems. From our perspective, the process of socioecological change is continuous and leads to emergent food system structures that will not remain equal until a new period of transition begins. Human societies evolve side by side with nature; however, by its own factors and mechanisms. This recognition of the essential unity of the evolutionary process implies conceiving social change as that in which the new emerges from the old, innovation elaborates on preexisting material. It was Edgar Morin (2010) who suggested that the necessary change toward a more sustainable world is a process of metamorphosis, a new qualitatively different socioecological order that must be built, however, on the existing foundations (see Chapter 5). By doing so, it distances itself from the eternal contradiction between reform and revolution, between evolution and rupture. Metamorphosis is thus an appropriate metaphor to understand the enormous complexity of socioecological change. Social systems do not evolve linearly but in unexpected, random directions, among other things, because they are entropically undetermined. Ultimately, we understand the food regime transition as the temporal process in which the most relevant changes take place leading to one of the isomorphic models of the food system we are considering. In that sense, the concepts of socioecological change and metamorphosis complement each other well, allowing transition to be understood as a process by which the food regime changes, for example, from an industrial to a sustainable mode. Metamorphosis admits variably lasting hybrid forms in which the food regime is neither entirely industrial nor organic.

The analysis of the driving forces of transition is also a complex task. It consists of exploring how the food regime's material processes (appropriation, circulation, transformation, consumption, and excretion) function mediated by a combination of intangible factors (knowledge, technology, institutions, etc.), and how that function changes as a consequence of the differential role played by the system's components through time. This process is directly linked to the complex, intricate, and reciprocal interrelations proper to the dynamics of ecosystems and landscapes being appropriated, or in which wastes are disposed of. Lachman (2013, 274) was right to criticize this approach to the socioecological transition: he claimed that the sociometabolic transition proposal constitutes an exceedingly abstract system (social metabolism) from which social actors are excluded. Factors such as beliefs as well as political, economic, or cultural interests are not accounted for, which makes it unlikely that this general and abstract model can guide its users in the design of policies to advance the transition. Our proposal considers that social actors are unquestionable protagonists. With the latter assumption in mind, collective action – collective action by agroecological movements in particular – is assigned a major role in the process of food regime transition. In previous epigraphs, we approached their negentropic capacity to promote changes that would make the complete metamorphosis of the food regime possible. In the following chapters, we attempt to diagnose the workings of the current food regime. We will also develop strategies and instruments of action that help agroecological movements to change the regime.

2 The Industrialization of Agriculture and the Enlargement of the Food Chain

With the industrialization of agriculture, agroecosystems have undergone a fundamental change: they have been expelled from the energy system and have become a recipient of energy and materials from elsewhere. Agriculture went from being at the heart of the metabolic process to constituting one of its apparently marginal segments, thanks to the exploitation of fossil fuels. This metamorphosis, which occurred at an accelerating pace, began in England, made the leap to continental Europe, expanded toward its peripheries, was taken to the colonies, and today is still spreading to every corner of the globe. The domestic extraction of biomass represented between 95% and 100% of the energy consumption in organic metabolic regimes, whereas in most developed societies where the industrial metabolism has become the dominant way of organizing relations with nature, biomass only produces between 10% and 30% of consumed energy. Furthermore, energy balances show that agriculture has changed from being a supplier to a demander of energy (Leach, 1976; Pimentel & Pimentel, 1979; González de Molina & Guzmán Casado, 2006; Tello et al., 2014; Guzmán Casado & González de Molina, 2017). Without the subsidy of external energy, a part of global agriculture could not function.

Agrarian activities have also changed their metabolic functionality. The socioecological transitions to the industrial metabolism regime have been accompanied by an accelerated increase in the consumption of materials, both in absolute and per-capita terms, especially from abiotic materials during the second half of the twentieth century. On a global scale, it was in the late 1950s when the extraction of abiotic materials came to exceed the extraction of biomass (Kraussman et al., 2009, 2011; Gierlinger & Kraussmann, 2012; Singh et al., 2012; Infante et al., 2015). In spite of this, biomass has gone from being the main source of energy and materials to specializing in two essential functions: the supply of food and the provision of raw materials for industry, especially wood, which is difficult to substitute in many industrial processes (Infante et al., 2014b; Infante & Iriarte, 2017).

Commercial crops per land unit have multiplied, offering the capability of feeding a population that has grown sixfold since the start of the nineteenth century. These changes have been possible as a result of the growing application of abiotic inputs into agrarian production: fossil fuels for machinery and irrigation, chemical fertilizers, and pesticides (Infante et al., 2014a). According to Smil (2001), the total area of farmland in the world grew by a third during the twentieth century; however, because productivity has multiplied by four, the harvests obtained in this period multiplied by six. However, as Smil himself acknowledges, this gain is partly due to the fact that the amount of energy used in farming is eight times larger (2001, 256). It also, and particularly, explains the exponential growth registered in terms of the productivity of agrarian labor. The case studies conducted for Austria by Krausmann et al. (2003) and Spain (González de Molina & Guzmán Casado, 2006; González de Molina et al., 2019) mostly concur that the industrialization of the agrarian metabolism led to a spectacular increase in the productivity of labor, due to the mass use of new technologies and the mass input of external energy.

The food market has gone global, forcing agricultural products to travel long distances before reaching the consumer's table, and even when eaten fresh, they require huge logistics infrastructures. Processed food has taken over fresh food and the amount of meals eaten outside of the home increases day by day. New activities between production and consumption, unheard of in the past, have come to the fore and acquired paramount importance: food transformation and delivery. New sophisticated appliances, using gas or electricity, take part in human nutrition and have increased the energy cost of food (Infante-Amate & González de Molina, 2013; Infante-Amate et al., 2018a), but at the same time, the food regime that rules the world's food system is incapable today of nourishing the whole of humankind (Dixon et al., 2001, 2) – even though there is enough raw harvest to do so. Very scant progress has been made to eradicate rural poverty. Moreover, the way food is eaten nowadays causes not only huge health problems to humans but unhealthy agricultural systems as well, including those of developing countries.

In short, the industrialization of agriculture and food system has converted industrially managed agroecosystems into highly entropic dissipative structures that require large amounts of external energy and materials, transferring the generated entropy to the environment and, consequently, having a big impact on it. This reflects a fundamental change: agriculture has become a business, a source of capital accumulation rather than a sustainable way of satisfying the food needs of the human species. This kind of agriculture is not sustainable and may therefore lead to collapse. The risk of collapse has increased considerably in recent years, and seems evermore likely if the food regime promoting this model of agriculture does not change. We will now address the origins of industrial agriculture in the following sections, reviewing its history (in this chapter) and its latest trends (Chapter 3).

2.1 THE ORIGINS OF INDUSTRIAL AGRICULTURE

Understanding how the transition to the industrial metabolic regime took place in agriculture is not a vain academic exercise. From a Political Agroecology perspective, it presents a special interest: it provides valuable knowledge to design the agroecological transition. The industrialization process of agriculture came about in three major waves: the first was fostered by an institutional change toward capitalism and took place within the boundaries of the agrarian sector, leading to the optimization of its capacity by raising biomass production; the second wave was the first metamorphosis of the sector due to the use of artificial fertilizers, that is to say, through the injection of energy and materials from non-renewable sources which effectively constituted an external subsidy; and finally, the third phase was the total penetration of fossil fuels within the agrarian sector (Krausmann et al., 2008b).

These three waves took place in an institutional context that contributed decisively to the industrialization process. It can thus be held responsible for its social and ecological impacts. This institutional context corresponds to what Marx understood as capitalism, although it cannot be reduced to capitalism in view of how the so-called socialist regimes unfolded in the twentieth century. In any case, given the episodic nature of such regimes, which in some cases represent extreme forms of state capitalism, we will refer to the basic institutional framework that has prevailed since the industrial revolution until today. Marx called this set of rules and norms "social relations of production", although their scope goes beyond production, extending to the distribution and consumption of goods, to all kinds of economic activity. These social relations shape a social architecture, and although this architecture may be abstract, it is nonetheless real and lasting: it delimits the spectrum of possible economic interactions between individuals (Wright, 2005, 590). As we saw in Chapter 1, social relations have the function of regulating not only social entropy but also the trade-off between metabolic entropy and social entropy. Together with technological and environmental limitations, these relations determine the context in which the action of peasants and farmers, in our case, takes place.

After a long period of marked fluctuations and scarce development, the European population began a general growth trend since 1750. In less than a century the number of inhabitants in the region duplicated (Wrigley, 1985, 2004; Livi-Bacci, 1999). At the same time, during the onset of industrialization, agriculture played the role of a determinant in physical terms. Industry required large amounts of labor and animal power, and the sustentation of a growing non-agrarian population. This was occurring before the new technologies could provide energy or material inputs outside the agrarian sector. British agriculture, for example, continued to be organic, and it would remain so for a long time. The main challenge was then to raise production without depleting soil fertility. During that early stage of British industrialization, as stated by Krausmann et al. (2008b), satisfying growing demand became a narrow bottleneck for economic growth. Indeed, population and consumption growth exerted pressure on forests, favoring the expansion of agricultural land for producing foodstuffs. In an opposite direction, households' growing need for fuel also exerted pressure. The increasingly widespread replacement of human labor by animal power (Wrigley, 1993) threatened neutralizing the improvements obtained from the agrarian revolution and increasing the costs of animal energy. All these requirements converged in creating a disproportionate demand for land impossible to assume by the British territory.

The agroecosystems could meet requirements by implementing one or several of the following four possibilities depending on their provision of land and their climate and soil conditions: (i) pushing the agricultural boundary as far as possible; (ii) saving land, increasing yield per land unit; (iii) specializing production, promoting one land use over others; and (iv) importing biomass that agroecosystems are unable to produce. Undoubtedly, freeing up the energy functions of fallow and pastureland

through the introduction of coal in economic activity, and even in domestic consumption, facilitated the reclamation of lands and their use for crop cultivation.
Farmed land area grew by 58% and the area of land dedicated to cereals by 62.8%
between 1700 and 1830 (Schandl & Krausmann 2007, 87). From this perspective,
the growth possibilities of agricultural production in the UK depended not only on
the innovations of the agricultural revolution but also on breaking away from the
rigidity of the feudal institutional framework and, therefore, on the possibility of
having access to more land to meet the endosomatic consumption of its population,
particularly in urban areas.

The most well-known innovations in saving land and increasing yield per land unit
took place precisely in the UK and gave rise to the so-called *Agricultural Revolution*
that, in the opinion of almost all historians, sustained the Industrial Revolution. The
latest historiographical contributions, however, do not talk about sharp changes but
rather of the slow introduction of improvements in the eighteenth century that boosted
productivity (Overton, 1991; Allen, 2004; Warde, 2009). New rotations, combining
cereals with leguminous crops and fodder, allowed for a better association between
crop and livestock farming, the increase in livestock numbers, the substitution of
human labor with animal traction, and an increased availability of manure, practically eliminating fallow. With the increase of production, farm income also grew,
and productivity also improved by releasing manpower that could be employed by
industry feeding at the same time a growing urban population.

According to Krausmann et al. (2008b, 194), the Austrian solution was similar to
that of Britain. These innovations had spread to Denmark, Hannover, and to some
other regions of Germany and Switzerland by the end of 1700, and during the early
nineteenth century they had reached France and other countries. The increase in biomass production was an indispensable condition for the advancement of the urbanization process, as can be seen in the territory needed for supplying basic foodstuffs
and other raw materials of agrarian origin to the rapidly growing European cities of
that time. The studies carried out in England, the canton of Berne (Pfister, 1990), certain areas of Andalusia (González de Molina et al., 2010) and Catalonia (Tello et al.,
2010) show that marketable production generally increased through the promotion
of agricultural crops and productive specialization which tended to break the balance
between the different land uses that had characterized organic metabolism.

The fourth solution was to turn to the markets to import nutrients and foodstuffs
for humans or animals. In fact, the growing demand of biomass for feeding humans
and animals was one of the main drivers of the *first globalization*. In around 1870,
when the potential for modernization within the solar energy-based agrarian system
had been exhausted, the UK changed its economic strategy by importing growing
quantities of basic foodstuffs from other parts of the world (and by adjusting its surplus population via emigration). Alf Hornborg (2007, 268) estimated that in the UK,
in 1850, the exchange of manufactured cotton fabrics for wheat, which obviously
did not need to be grown inside its borders, represented the sparing of nearly three
million hectares of cultivated land that could be assigned to crops other than wheat.
In 1900, the land area equivalent to the imported cereals achieved a similar extension to that available for domestic farmlands (Krausmann et al., 2008a, 194). This
flow of cereals even permitted the British agrarian sector to specialize in livestock

production, which consumes a great deal of land but saves on manual labor and produced lean benefits for large English landowners. However, this appropriation of more land was not always peaceful; on a number of occasions, it was achieved with political–military means. Nineteenth-century colonialism is a good example of this.

Consequently, the industrialization of Europe was aided by trading with many peripheral countries; most of them net biomass exporters. The supplies of agrarian products (coffee, cocoa, sugar, cereals, fruits, livestock, produce, etc.) was in practice a territorial subsidy provided by peripheral countries at a low cost (Schor, 2005), which gave Europeans more available land for specialized production. The same happened with fertilizers and other agrarian inputs imported from poor countries. Poor countries exported their soil nutrient reserves (Mayumi, 1991) as crops of sugar cane, cotton, etc. The cases of the Peruvian guano and the Chilean saltpeter illustrate well the effects of these uneven exchanges, not because of the actual impact it had on European agriculture, but because of the net loss of productive capacity (Leff, 1986) implied in the exportation of these resources.

2.2   THE INSTITUTIONAL FRAMEWORK: PRIVATE PROPERTY AND MARKET

These four solutions to meet growing biomass demand were confronted with the legal–political structure that protected the traditional configuration of the organic metabolic regime and particularly the distribution of land uses. This contradiction facilitated institutional change (by liberal revolutions, but not only) especially to the regime of feudal ownership and the "liberalization" of agrarian markets. The new liberal regimes paved the way for markets, individual freedom, private property, and free trade. The long-standing debate between protectionism and *freetradeism*, with its diverse and complex arguments in favor or against either of the two options, can be better understood from the material perspective. Countries such as the UK chose free trade because among other things, the country was unable to sustain the urban demand of its industrialization process with its domestic resources. In Spain, the opposite occurred. Croplands were available, urban demand was not huge, and protectionist policies were present throughout most of the nineteenth century.

In any case, the legal changes coming with the building of liberal regimes and the configuration of nation states instituted private property and markets – first in Europe and later, on all the continents. Both established – and still form – the two main pillars supporting the allocation of goods and services as well as the access to the resources necessary to produce them. The fact that mercantile relations turned into the dominant relations, that is, into the preferential way of assigning goods and services, had significant consequences. In reality, a market is an instrument used to transfer energy and materials between different social groups or territories, which in turn consume energy and materials and produce waste. A society can increase the carrying capacity of its territory by importing resources from other societies through economic exchanges. From a biophysical point of view, the formation of markets in agriculture was also favored by the need to overcome certain production factors (water, traction, and nutrients) that limited production growth. Production commodification and the "emancipation" of the limits imposed by the agroecosystem's land provision (and the quality of its soil and climate) are closely linked: energy and materials subsidies circulated through the market itself, the market was thus crucial to sustain the growth of agrarian production. At first, the sphere was the local district, then the province, and later it became regional, national and finally international. The spatial scale of commodification gradually increased until it would become, as it is today, a global phenomenon. But the market also circulated requirements and pressures from other territories, via prices, on local agroecosystems, generating demands in excess of the real demands of their population. The drive to obtain sufficient income to survive or to maximize profits or earnings constituted a powerful mechanism that was increasingly expressed in the market and pushed toward metabolic change in agriculture (Berstein, 1977, 1986; Harris, 1982; Marsden, 1991).

For its part, the introduction of private property in terms of absolute ownership over the property in one's hands, in this case the land, excluded any collective or common lands. Different forms of communal governance of both natural and social resources were thus set to disappear. The consequences are well known: peasants were dispossessed of rights and common lands, leading them to depend almost exclusively on their farms to subsist. Those with no access to farms were condemned to either poverty or the labor market and low wages. Yet there were major consequences also from an agroecological viewpoint: these have received scant attention, despite their enormous relevance regarding the viability of sustainable agriculture. The response to growing human food demands led agriculture to progress at the expense of pasture and forest lands, dismantling the integration between different land uses; conversely, it would also increase the amount of land dedicated to animal feed to the detriment of the other two. The fact is, the closing of biogeochemical cycles, made possible thanks to agrosilvopastoral integration, became impossible in the new scenario: the ownership of a piece of land dedicated to agriculture or livestock made it very difficult to close such cycles. For example, in Spain, communal property had the basic function of feeding the livestock that provided the manure necessary to fertilize the most intensive farms. The privatization of communal property meant that peasants resorted to markets to obtain the necessary food for their livestock or the manure needed to replenish the fertility of their farms. In this sense, private property constituted a powerful mechanism to commodify production, creating an input market that would eventually become a market for inputs outside the agrarian sector. Private companies increasingly participated in input markets, and often headed technological innovation; inputs became definitively incorporated into the market and from then on, ceased to be possible to obtain from the farms themselves. This meant that it became impossible to intensify production independently and autonomously, without relying on markets (Ploeg, 1993). Finally, private property established a stable form of social inequality in the countryside; lacking any preventive "cushion", it divided large and small producers, landowners and landless peasants. We will address the sociometabolic consequences of this new societal configuration later.

The replenishment of soil fertility became the crucial factor. In many countries the expansion of the agricultural boundary, or the increase in yields to meet those growing demands, was achieved thanks to changes in land uses and a better integration of livestock in agriculture, but it also came at the expense of nutrient reserves accumulated in the soil over centuries either naturally or through management of farmers (Garrabou & González de Molina, 2010; González de Molina et al., 2010). Throughout the nineteenth century, however, the possibilities of maintaining sustained growth of agrarian production through the progression of agricultural cultivation, or through the increase in the land's productivity, were exhausted. In some countries this occurred early on, in others, it came later, but most European agroecosystems had reached their productive limits by the end of the nineteenth century. Technologies associated with coal were not yet able to supply external energy to cultivation, only the use of coal alleviated the pressure on forests. Agriculture would continue to have an organic energy base even in countries adopting industrialization. The relative scarcity of nutrients, exacerbated by failed territorial equilibrium due to production growth during the nineteenth century in Europe, is one of the major reasons that caused the end-of-century crisis. It initiated the second wave of the agrarian socioecological transition.

2.3 THE FIRST GLOBALIZATION: THE EMERGENCE OF FOOD REGIMES

Tariff liberalization not only favored the worldwide expansion of private property and the market, it also converted the liberal economic order into an international regulator of economic and physical biomass exchanges. The *first food regime* extends from the 1870s to the First World War and corresponds to the so-called *first globalization*, built on the basis of an international cereals market. It was driven by Great Britain thanks to its superior military power, the free trade discourse, and the pound sterling as the key currency for international trade (Dörr et al., 2018, 182). However, the rapid industrialization of the USA, Germany, France, and Japan led to growing competition for the control of new emerging markets and the control of raw materials between these nations. That competition was a "game" played by colonial powers, colonies, and the recently established nation-state system (Bernstein, 2010, 41). European states imported large amounts of wheat and meat from settler states, who in return imported manufactured goods, labor, and technology, especially for railway construction (Friedmann & McMichael, 1989, 95). Fertile lands in the USA, Canada, Argentina, or Brazil, usually violently appropriated, were devoted to the production of grains and meat at lower prices, subsidizing food to European workers. In the USA, the exports of cheap grains were possible thanks to soil fertility accumulating year by year in the territories that just colonized by the new settlers. As Cunfer and Krausmann (2009) have shown, the grain production in the Great Plains produced soil mining. A vast amount of land was also devoted in Asian and African colonies to produce what Bernstein (2010, 68) called "tropical groceries" – sugar, cocoa, coffee, tea, bananas, and other crops for European consumers. By this means, industrial countries ensured a cheap food supply with which to lower, in turn, the reproduction cost of the working classes, that is to say, the wages of the industrial sector (Moore, 2015). Cheap food also gave political stability to the liberal governments at a moment of strong workers' protests and social unrest.

This newly established global market for grains led to the emergence of productive specialization in dairy, fruits, and vegetables. Indeed, it had triggered, in return, an agrarian crisis of great proportions, called a *fin-de-siècle* agrarian crisis. Undoubtedly, cheaper grain prices stimulated the productive specialization of European farmers and probably reduced their income, favoring productive intensification strategies to compensate for this reduction. As we have seen, this specialization was possible thanks to the first chemical fertilizers. Indeed, the successive land arrangements, designed in the nineteenth century to produce new essential balances, became expensive and impracticable owing to their growing size. The continual increase of agricultural surface area and its productive intensification aggravated the nutrient deficit to such an extent that it cost ever more money and efforts to cover this deficit by importing organic fertilizers. This created a favorable context for the spread of land-saving technologies, especially chemical fertilizers. More intensive rotations, without fallow and with successions of crops that would have been previously impossible, were now possible, stimulated by the integration of international agrarian product markets at the end of the nineteenth century. Around 1904, the German chemist Fritz Haber began to experiment with the possibility of synthesizing ammonium, a form of reactive nitrogen, the shortage of which in soils poses a strong limitation to agricultural land productivity (Smil, 2001). Chemical nitrogen fertilizers and other agrochemical contributions, and later advances in genetics, contributed during the early twentieth century to significantly transforming crop fields.

2.4   GREEN REVOLUTION AND THE SECOND FOOD REGIME

Following a long transition period, a *second food regime* (1945–1973) was established after the Second World War, now under the USA's political, economic, and monetary hegemony. However, this hegemony was countered by the other pole of the Cold War and the block policy, the Soviet Union, which actively sought new allies among the countries emerging from decolonization. The main actors of this new international order thus continued to be the nation states and the US dollar – the international trade currency. International markets – including international food aid – not only turned into outlets for surpluses, mainly for cereals, generated by national production intensification policies, but also led to opening markets for large-input companies. The power of the USA prevented international free trade agreements (GATT) from applying to the food sector and favored the adoption of its export subsidy policy by most developed countries. National policies encouraged production intensification and specialization, bolstering agricultural industrialization. It led the agricultural sector to becoming subordinated to industry and services and the basis of capital accumulation. Its role consisted in providing cheap food, transferring resources to develop other economic sectors, and serving as a vast market for the input industry.

   Indeed, the possibility given by oil and its associated technologies of injecting large amounts of energy and materials radically changed the world's agricultural scenario (Krausmann et al., 2008b). The main transformations took place after the Second World War in the form of the Green Revolution consisting of improved seeds, chemical fertilizers, mechanical traction, and pesticides. Associated with them, two basic innovations for the industrialization of agriculture permitted the mass subsidization of agriculture with external energy: electricity and the internal combustion engine. It started with the mechanization of many agricultural tasks and, in most rich

TABLE 2.1

Indicators of Agricultural Industrialization in the World

	Units	1963	1978	1993	2008
Fertilizers (N)	1,000 t	15,011	53,327	74,493	105,738
Rural population	Millions	2,106	2,656	3,134	3,385
Mechanization	1,000 tractors	12,389	20,557	26,003	–
Cereal yields	kg/ha	1,321	1,946	2,502	3,149
Food energy	Petajoules	11,027	16,075	22,393	29,060
Energy intake per capita	Kcal/day	2,253	2,451	2,636	2,822

Source: FAOSTAT and González de Molina & Toledo (2014).

countries, culminated with the spread of the Green Revolution technological package at the end of the 1950s (see Table 2.1). The presence of animal traction impeded further expansion of agriculture and intensive livestock farming. It was necessary to develop a kind of technology that would once again save land, freeing up working livestock productive areas, and a kind of technology that would replace animal traction with mechanical traction. Added to that was the convenience of saving costs to achieve a minimum threshold of profitability, situated at a lower level than the average profitability of other economic activities. The reduction of manual labor, replaced by machines or by chemical means that made certain tasks easier (weeding, for example), was the solution. In some countries, emigration from the countryside to the city and the development of movements of paid farm laborers pushed wages up and sped up the substitution process.

Many countries adopted a policy of import substitution (for a review, see Bruton, 1998), financed by the agrarian sector that was experiencing a new intensification process. This was the goal sought by the modernization policy that accompanied the Green Revolution in peripheral countries. Cultivated land areas expanded, par-ticularly those dedicated to commercial crops. Permanent grassland also extended at the expense of forested areas. Between 1970 and 1985 alone, the forest surface area in Latin America and the Caribbean region fell by nearly one million square kilometers. The expansion of cropping and irrigated land unsuitable for agriculture also had severe consequences. The FAO map of soils in 1990 included 400 million hectares of degraded soils in Latin America alone (GLASOD, 1991).

New crops depended on improved seed varieties, needed large doses of fertilizers and pesticides, and required agricultural technology beyond the reach of poor coun-tries. In 1984, 20 times more fertilizers and 25 times more pesticides were used in Latin America than in 1950. From 1950 to 1972, the annual rate of average consump-tion of fertilizers grew by 14%. By 1980, Latin America was spending $1.2 billion on pesticides (FAO, 2016). Hunger, poverty, and malnutrition did not disappear, but technological dependence and debt grew in an unusual way. In fact, the translation of the Western model of intensive agriculture to countries with different edaphic and climatic conditions opened up a huge market for transnational agrochemical and food corporations. Perhaps an extreme example is what happened in Brazil with the use of pesticides. Pesticide consumption in Brazilian agriculture went from 599.5 million liters in 2002 to 852.8 million in 2011. Since 2008, Brazil has become the largest consumer of pesticides in the world. The quantity used per unit of surface went from 10.5 liters per hectare in 2002 to 12 liters in 2011. In 2011, the cost of applications reached US$ 8.5 billion (Carneiro et al., 2015, 50–52).

In parallel, a major destruction of the peasant sector was taking place, conditioning the loss of food self-sufficiency (Toledo et al., 1985). The expansion of the livestock industry during the post war period is a good example of such a phenomenon. As standards of living rose in industrialized countries, the consumption of animal protein also increased, in particular that of meat and dairy products. In order to supply the continually growing demand, peripheral zones were devoted to raising livestock or producing fodder. Global meat exports grew from 2 million tons in 1950 to 11 million tons in 1984. Many countries in Africa and Latin America converted extensive cropping areas to grazing land for cattle. In particular, large areas of forest land were converted to grasslands, while in other countries traditional crop varieties were replaced by a monoculture of forage crops. In both cases, the result was a growing production deficit of cereals and other foodstuffs formerly grown domestically (for a review, see Barkin et al., 1991). In addition, the modernization of agriculture was achieved at the expense of traditional farmers, who were forced to migrate to cities to live in conditions of extreme poverty or remain farming marginal land.

Those who remained farmers enjoyed no better living conditions. The biased distribution of property, the trend to concentrate land tenure in a few hands, and the destructuralization of rural communities brought about by modernization forced farmers to cultivate forested and marginal lands. Many deforestation processes, overgrazing, cultivation of slopes that in some zones have accelerated erosion and desertification – Sahel, India, Panama, Brazil, etc. – are associated with that practice. However, perhaps the most decisive change, owing to its impact on the species itself, has been the change in diet. Rich countries increasingly consume more meat and livestock products such as milk and its derivatives, causing livestock numbers to grow to surprising levels. To feed these animals, land has been taken away from growing food for human consumption, or part of it has been dedicated to growing feed to fatten livestock. According to Krausmann et al. (2008a, 471), the global appropriation of land biomass in the year 2000 reached 18,700 million tons of dry matter per year, 16% of the world's net primary production, of which 6,600 million tons were indirect flows. Of this amount, only 12% of the vegetable biomass went directly toward human food; 58% was used to feed livestock, a further 20% as raw material for industry, and the remaining 10% continued to be used as fuel.

The importance acquired by imports of energy and materials have led agriculture to become partially decoupled from the ecosystems that sustain it, and its spatial configuration to become radically different, being based on simplified landscapes, single crops, the loss of spatial heterogeneity and biodiversity. Basic functions that had been formerly fulfilled by the land (production of fuels, food for livestock, basic foodstuffs for the human diet, etc.), and to which a fairly large portion of land was dedicated, have disappeared, giving rise to a specialized landscape, peppered with constructions and areas used for urban–industrial properties (Agnoletti, 2006; Cussó et al., 2006; González de Molina & Guzmán Casado, 2006; Tello et al., 2008; Guzmán Casado & González de Molina, 2009).

Contrary to what happened during the first food regime empire, the second was characterized by basic grain exports from the North, and exotic exports from the South. The USA, Canada, Australia, and later Europe, were the main players in the new world food market where production surpluses found outlets: the so-called developing countries, who followed the same economic policy of providing cheap food to foster the national industry. As a result of this misguided policy, prices in domestic markets sank, many countries lost their food self-sufficiency and became net food importers (see the case of Mexico in Toledo, 1985). Meanwhile, they exported raw materials destined to northern food industries, facilitating the generalization of the consumption of meat, dairy, and processed food and favoring diet change and the so-called "nutritional transition". The "modernization" of the sector, i.e., production intensification and specialization, became a solution for both developing countries and the most industrialized countries.

2.5 THE MAIN DRIVERS OF AGRICULTURAL INDUSTRIALIZATION

But what were the factors that drove these fundamental changes? The answer is highly relevant and allows us to understand the deep-seated mechanisms, that is, to know the industrialization processes, in our quest to pilot a new transition toward sustainable agriculture. The literature tends to consider the specialization of agricultural production, encouraged by the market, as the main driver of the industrialization process (for a review see IPES-Food, 2016). But this analysis overlooks another major driver: production intensification, responsible for most of the environmental, economic, and social problems caused by industrial agriculture. By emphasizing productive specialization only, and ignoring the process of intensification, we only identify part of the problem and, therefore, only part of the solution (specialization versus diversification). In other words, two main drivers have pushed the industrialization process forward: productive specialization, fostered by integrations of agricultural markets, and intensification, caused by the continuous loss of profitability of agricultural activity.

Specialization and intensification, in addition to being complementary, have been synergistic. Both processes have their roots in social inequality fostered by capitalism, playing a key role in the agriculture industrialization process. To better understand this role, it is useful to distinguish between two types of inequality: internal and external inequality (Guzmán et al., 2000). Indeed, the unequal distribution of agrarian income put direct pressure on farmers to compensate for declining income by producing more (intensification) or specializing production into crops or livestock breeds with greater added value or easier access to the market (specialization).

From the perspective of agriculture's internal equity, unequal distribution of agrarian resources creates pressure toward a greater productive endeavor, giving rise to different strategies among large landowners. For example, liberal regimes increased unequal access to land in Europe and placed peasants in a very precarious position: the legal–political framework of the feudal or tithe-based system, which guaranteed the existence of common goods and rights that attenuated poverty, disappeared. Peasants were thus pushed toward new relationships of a more mercantile and monetary nature. This process of peasant dispossession has continued to occur worldwide under the different food regimes, with very similar consequences. Peasants are forced to increase productivity per hectare through more intensive crops; this would provide them with sufficient income to cover the payment of higher rents, or to buy basic goods for subsistence, many of which they could no longer obtain free from common goods or rights of which they had been stripped. Furthermore, as land became an object of private appropriation, the fragmentation of farms and their decreasing size obliged peasant families to increase their productive effort; they also usually searched for alternate sources of income in the labor market.

For their part, large landowners or farmers develop productive specialization strategies to secure higher incomes. Large owners adapted to the new European liberal legal framework, redefining their strategies of promotion or maintenance of their social status. The landowning classes wove patrimonial strategies relating to the land, among which inheritance and homogamic marriage were privileged instruments. To support such strategies, and the lifestyle and consumption imposed by the bourgeois *habitus* of the time (Bourdieu, 1979), they had to achieve maximum profitability from their possessions, which was done by increasing the income received from farmers, accumulating more wealth, but also increasing yields (of the most commercially oriented plants or livestock) per unit area.

However, low agricultural income can be caused not only by unequal access to fund assets (land, livestock, and technical capital), but also by deteriorating exchange relationships between the agrarian sector and the urban–industrial sectors of the economy. Agrarian incomes thus become lower than incomes of other sectors of activity, in other words, the cause can be external inequality. The subsidiary role given to the agricultural sector, devoted to the supply of cheap food, was usually translated into low prices perceived by farmers. The global profitability of agrarian activity, even though it was still linked to crops that were in higher demand and therefore more profitable, declined progressively from the early twentieth century onwards as a consequence of the unequal relationship of exchange between the agrarian sector and the industrial and service sectors (FAO, 2004). The terms of exchange for primary producers declined by 1% per year between 1948 and 1986. Although the evolution between 1900 and 1998 was not uniform, but rather developed in stages, with two collapses in 1920 and 1984, the accumulative effect was a decline of 62% in the terms of exchange (Zanias, 2005; Eisenmenger et al., 2007, 183). The economic policies implemented by governments coincided in favoring this unequal exchange, through their endeavor to promote industrial development.

The loss of profitability fostered a new process of crop intensification with which farmers and large landowners attempted to offset diminishing agrarian income. This guided the choice of crops and management techniques implemented by large and medium-sized farmers, familiarized early on with the market, and represented another turn of the screw for productive specialization. They also adopted technological improvements that, supported by a thriving chemical and mechanical industrial sector, allowed them to boost yields and productivity (Koning, 1994). Far from recovering lost profitability, this process made farmers even more dependent on the market and new technologies, in other words on the agroindustrial complex as a whole, to achieve a minimum income threshold. Farmers were trapped, doubly squeezed by input prices and retail market prices (Owen, 1966; Marsden, 2003). Until this vicious circle became generalized, inequality originated in the allocation of land and livestock ownership, but as it became necessary to acquire essential factors of production through the market, the market became the fundamental mechanism of exploitation of agricultural labor (Bernstein, 1977, 1986, 2001) and of transfer of part of agriculture's value added to other sectors of activity: the input industry, large retail and distribution or the food industry.

In short, this non-egalitarian relationship between the agrarian sector and the rest of the economy occurred mainly through the combination of two closely related phenomena: on the one hand, the constant fall in prices perceived in real terms by farmers, a trend that remained constant practically throughout the twentieth century (see, for example, part III of the FAO, 1995). On the other hand, the growing use of industrial inputs, which incorporated greater added value, and with which farmers attempted to increase their revenue in order to counter the fall in perceived prices, led them to incur ever more onerous costs, ultimately reducing their net margin. In short, the deterioration in agrarian income created a favorable context for the rapid spread and mass use of agricultural inputs. The growing insufficiency of earnings became a powerful driver of agrarian intensification and of small farmers' engagement with markets through the purchase of inputs.

Intensification and commodification were, therefore, two closely linked processes which spiraled once they had a grip on production. At one point in their development, they generated a demand for energy and materials from the agroecosystem that the latter could not fulfill, exceeding its capacity for sustainment and degrading its resources and environmental function. Environmental deterioration created a new obstacle to intensification which had to be overcome through new technological changes; technologies that required, in turn, a more intensive use of energy and materials that eventually deteriorated the agroecosystem even more, spreading environmental degradation to other territories. Cheap fossil energy allowed agriculture greater access to mechanization, as well as to chemical fertilizers and pesticides, thereby facilitating industrial modes of production (UNEP, 2012).

In sum, the agricultural sector, as prescribed by economists since the Second World War, should always support the development of a country by providing capital, labor, a market for industrial inputs and cheap food. Most economic policies in the world have tried to control and reduce the value of wages in industry and services as a way to promote the development of economic activities with higher added value, pushing agricultural prices down. Since the beginning of the Industrial Revolution, and more conspicuously in the last century, agriculture has been an "extractive" sector whose mission has been to provide cheap food and raw materials to foster economic growth based on industrial production and economic services. The agrarian sector has been tasked with compensating for the social entropy caused by implementing an economic model based on social inequality and the accumulation of capital, but the mission has come at the expense of converting agriculture into an activity of high biophysical entropy. The institutional framework described in the previous section, based on private property and the market, reinforced by nation states' economic policies, has created the context that has made this possible.

2.6   BEYOND AGRICULTURE: THE FOOD SYSTEM

The industrialization of agriculture has also been accompanied by a fundamental change in the ways in which we consume food (Tilman & Clark, 2014; Steffen et al., 2015a), requiring ever more energy investments. The business it has turned into has given rise to new economic activities that intervene between agricultural production and the end consumption of food. The gap has grown over last decades, turning food into an increasingly industrial product resulting from the transformation of agricultural products and the addition of different utilities (Lancaster, 1966). As claimed by Dörr et al. (2018, 186), power shifted to input corporations – which provided green revolution technologies – as well as to brand food manufacturers. The industrialization of food production opened new accumulation dynamics for the manufacturing and retailing of processed foods.

In the 1970s, two papers were published in *Science* analyzing energy consumption in the US food system as a whole (Hirst, 1974; Steinhart & Steinhart, 1974). They exposed how, in a short period of time, not only did the agricultural production witness dramatic transformations but also the rest of the food system. Energy use in food-related industries, commerce, and household activities increased fourfold between 1940 and 1970. Recently, some articles have studied energy use in the current food system in the USA (Heller & Keoleian, 2000; Canning et al., 2010) and Spain (Infante-Amate & González de Molina, 2013; Infante-Amate et al., 2018a). They highlight that off-farm activities in modern food systems are responsible for between 70% and 80% of total energy use, emphasizing the need to pay attention to what happens away from the farm when studying energy demand in food-related activities. The case of transport is the most recurrent (e.g. Martinez et al., 2010). Under the idea of "food miles", estimations of transport-related energy use in food systems have multiplied, generating ongoing debates about the relationship between the distance travelled per product and energy consumption (e.g., Schlich & Fleissner, 2005; Martinez et al., 2010). A great number of studies have also estimated energy use throughout the entire food chain in specific products based on life cycle assessment (LCA) methodologies (e.g. Roy et al., 2009; Clune et al., 2017).

There are no studies that quantify the functioning of the food system on a global scale, despite the importance that international trade in food and large agroindustrial companies has acquired in recent decades. To a large extent, food systems, despite being strongly linked, are still organized at the national level. Their operation is, without a doubt, very similar. Let's see it in detail, following the Spanish case and quantifying the energy use of different links in the food chain in the long term. This exercise is remarkably useful to understand the extent to which the human food process is full of negative social and environmental impacts. We analyze three different benchmark years (1960, 1985, and 2010), covering the period of the country's major socioecological transformation (Infante-Amate et al., 2018a). Throughout this analysis, it is possible to estimate the energy weight of food system activities in the Spanish economy as a whole and how it has evolved over time. By estimating the weight of the food system's energy use, and identifying major hotspots of consumption, we can also better identify the goals of an agroecological strategy, reducing the energy profile of the whole food system.

Between 1960 and 2010, the population multiplied 1.5-fold, apparent food consumption by 1.9, final energy consumption by 4.4, GDP by 8.5, and domestic material consumption by 3.6 (Infante-Amate et al., 2014, 2015), while the energy demands of the food system multiplied by 10.2 over the same period. The growth of the food system represents a typical case in the context of the *Great Acceleration* that occurred in the second half of the twentieth century (Steffen et al., 2015a). Primary energy consumption (including indirect energy) in Spain's food system increased 10-fold from 1960 and 2010. During the first phase, between 1960 and 1985, the greatest increase in energy consumption was found in agriculture, because this was the period of major transformations of agricultural industrialization. In 1960, it was the largest consumer of energy, with 25.5%. It rose from 46.3 PJ to almost 350 PJ between 1960 and 1985; that is, a 7.5-fold increase. In that period, it increased from 25.5% to 35.0% of total food system consumption, responsible for 38.8% of new energy consumption. Transport was the second sector in importance. Its consumption multiplied by 6.1, and its share within food system sectors rose from 20.0% to 25.2%. The second period, from 1985 to 2010, presented different characteristics: the growth of agriculture slowed substantially and in relative terms, it was actually the sector to grow the least over the period, multiplying by 1.3. Energy demand in the transport sector doubled, becoming the highest energy consumer of the entire food system in 2010 (25.8%). In this second phase, home energy demand for food-related uses also grew significantly, multiplying by 2.5. Today, this latter demand represents the third largest source of energy consumption after transport and agriculture.

Similar figures could be appreciated in the case of the USA, the only country for which long-term evidence is available. Steinhart and Steinhart (1974) estimated that agriculture energy use within the whole food system rose from 18.2% to 24.2% between 1940 and 1970, a period during which US agriculture industrialized rapidly (Cleveland, 1995). Heller and Keoleian (2000) estimated that the share dropped to 19.4% in 2000. In the case of Spain, the share of energy used by the agricultural sector with respect to the whole food system has slightly dropped over the last three decades. Nevertheless, the evidence from these two studies shows that the industrialization of agriculture played a pivotal role in the initial stages of the transition and the problem of food's energy consumption cannot be limited only to the farm level: it has to be evaluated throughout the whole food system. Except regarding certain agrarian inputs from the 1980s onwards and the use of firewood for cooking between 1960 and 1985, the food system has grown, combining an increase in traditionally used inputs with the addition of new products.

But let's return to the case of Spain and briefly describe the current weight (2010) of the different links in the food chain. The industrial management of Spanish agroecosystems has led to sharp energy cost increases related to gas oils, electricity and, above all, to the production and transport of inputs required for agricultural and livestock production. Indeed, total energy consumed in farms (electricity and fuels) has increased to an astounding degree since 1960.[1] Agricultural machinery has grown

---

[1] According to IDAE data (Instituto para el Ahorro y la Diversificación Energética [Institute for Energy Saving and Diversification] (2013), *Balances del consumo de energía final*. Serie histórica: 1990–2011. Madrid: IDAE.), total energy consumption in agriculture doubled between 1990 and 2004, from 1667 ktep to 3324 ktep. The very dynamics of diminishing returns in the sector and the economic crisis itself have meant that since 2004, total energy consumption has fallen to 2,062 ktoe in 2010, returning to 1994 values.

TABLE 2.2

Total Primary Energy Consumption of the Spanish Food System, 1960–2010

Food Chain	Petajoules		Percentage		Increase	
	1960	2010	1960	2010	1960/1985	1960/2010
Agriculture	46.3	449.1	25.5	24.2	7.5	9.7
Transport	40.5	479.0	22.3	25.8	6.1	11.8
Industry	31.5	235.9	17.4	12.7	4.0	7.5
Packaging	4.9	190.4	2.7	10.3	20.3	45.1
Trade	14.5	192.9	8.0	10.4	3.7	13.3
Households	43.7	307.7	24.1	16.6	2.8	7.0
Total	181.4	1854.9	100.0	100.0	5.5	10.3

Source: Infante-Amate et al. (2018a, 572).

steadily while feed and seed consumption has risen ostensibly. The fuel and electricity requirements of fleets of tractors or irrigation pumps are substantial. Animal feed consumption rose from 5.9 Mt to 31 Mt between 1960 and 2008. That's not all. A key element of industrial agrarian systems is the artificial replacement of nutrients with inorganic sources external to the farm, accounting for almost 52,324 TJ, close to 12% of the agrarian sector's primary energy consumption. It is agriculture's third most important expenditure after the fuels used and the feed imported from outside the sector. The Spanish agrarian sector remains heavily dependent on grains from overseas, especially from Argentina, Brazil, and the US, used mainly for cattle feed.[2] This is how intensive livestock is maintained. It constitutes a major source of unsustainability, making it possible to mass-produce meats and dairy products at a low cost with major impacts on both Spanish agroecosystems and those of the countries mentioned above. The energy content of these grains accounts for almost 43% of the agricultural sector's energy consumption and 10% of the food system overall.

The agrarian sector therefore accounts for a consumption of almost 450 PJ, close to a quarter of the functioning food system's total primary energy consumption. The remaining two-thirds are used by the other links in the food chain, which have grown disproportionately, as shown in Table 2.2. The transport of all agrifood products in the domestic market is responsible for 25.8% of the primary energy consumed by the system as a whole, that is, 479 PJ, turning it into the link in the chain that consumes the most energy. A total of 78% of t-km of imported food is transported by sea and 20% by road and then carried by road to final destinations via a process that is far more costly in energy. This transport process justifies the "food miles" energy calculation measure. Today, Spanish transport accounts for 146,000 Mt-km, of which 33.6% goes to the transport of agrifood products. During the whole period under study, more than one of every five energy units consumed in the food system was dedicated to the domestic transport of food and agricultural products. Most transport was by road (almost 80% of the energy spent on transport and 20.3% of the food system as a whole), both for industrial and commercial transport and the transport of citizens traveling to big retail chains.

---

[2] Feed and cereals imports alone exceeded an average of 16 million tons over the last decade (AEAT, 2014).

The need to conserve food, which now travels long distances over long periods of time, requires packaging to keep it in good condition. Moreover, the abandoning of direct and local consumption as well as the disappearance of bulk, unpackaged purchases obliges points of sale to conserve and offer products in new containers that, in turn, form also part of new agrifood marketing. While the practice was non-existent or negligible in 1960, today 2.1 million tons of glass containers, 1.5 million plastic and 1.5 million paper/cardboard containers are consumed for food purposes. Bags currently used for purchases are somewhat less significant; nonetheless, they represent a similar amount to all the packaging consumed in 1960. Added to the environmental impacts derived from the use of these – often highly polluting – products, the energy they consume is barely inferior to the caloric content of the foods they contain.

The agrifood industry has also undergone a deep transformation over the last half century. The installed power of a large part the traditional industry was very reduced; it allowed processing only a small amount of food, but it concentrated and adopted more efficient transformation processes, in such a way that it now processes an increasing amount of food based on a smaller number of entities. The agrifood industry consumes 12.7% of the food system's primary energy requirements, the bulk of the spending on supplies coming from raw materials (98%). The distribution sector and food retail has also profoundly changed. The data reflect the disappearance of traditional trade and its replacement by new, large centers. The number of traditional shops dropped from 92,484 to 27,423 between 1988 and 2006 (Nielsen, in MAGRAMA, 2007). Self-services dropped from 17,893 to 10,305. In addition, the number of supermarkets multiplied by 2.7 and hypermarkets by 3.8, rising from 5,292 to 14,084, and from 99 to 379, respectively. In 1980, there were only 20 large shopping centers in Spain. In 2008, there were 514 (AECC, in Cuesta & Gutiérrez, 2010).

Another notable characteristic is that increasing amounts of food are consumed outside the home. In fact, in 2012, a third of Spain's total food expenditure was outside the home (MAPAMA, Food Consumption Panel in Spain, 2012). Between 1975 and 2010, the number of catering establishments, including cafes, bars and restaurants, rose from 156,000 to 328,000 (FEHR, 2005). Restaurants multiplied by 3.3, cafeterias by 3.5, and bars by 1.8. The number of hotels has skyrocketed, from 2,551 in 1960 to 16,938 in 2010. Major changes have also taken place in the way food is consumed in households. A total of 19% of total household energy consumption is related to food. The kitchen (40% of the consumption linked to food) and the refrigerator (36%) represent the biggest items, followed by ovens (10%), freezers (7%), and dishwashers (7%). Overall, household energy demand is one of the food processes to have grown the most, especially over the 1985–2010 period, multiplying by 2.5, from 123.8 to 307.7 PJ. The increase in meat product consumption, the abandonment of local and seasonal consumption, as well as the acquisition of products from other parts of the world require the use of domestic appliances for their conservation or cooking, which brings about a greater consumption of electricity and fuel. Food preparation and domestic conservation is today the third most energy-consuming activity after transportation and agricultural production.

The major transformation of the entire agrifood chain described above, characterized by increasing amounts of inputs and mechanized industrial processes, has generated a significant growth of energy demand, for both direct and indirect consumption, to produce the agrifood system's goods, buildings, or supplies. The energy consumption of the entire agrifood chain has increased 10.6-fold. It has risen much faster than total energy consumption, than the population, than the total consumption of food and even that of GDP. In short, food supply currently depends not only on the agricultural sector, but also to a large extent on industrial processing and transport. In total, more than 1,850 PJ is needed to satisfy Spain's endosomatic metabolism, while the energy contained in consumed food barely reaches 235 million GJ (Infante & González de Molina, 2010).[3] That is, according to cautious estimates, for each energy unit consumed in the form of food, six units have been spent on its production, distribution, transport, and preparation. The inefficiency of the human feeding process is a faithful reflection of the extent of its unsustainability.

2.7   A THIRD FOOD REGIME?

The possible existence of a *third food regime* is currently under discussion. Its existence depends on the degree to which each author formalizes the concept itself. Nevertheless, the regime's features are sufficiently prominent and relevant to be addressed in their own right; especially if we focus on the internal contradictions that threaten to make it collapse and that have dramatically intensified in recent decades. Given this predicament, the present status of the corporate food regime (CFR) should be considered beyond that of an "emergent regime", as described by Friedmann (2016). This does not mean, however, that, as this author criticizes, the regime is governed or coordinated by some consolidated supranational authority. In fact, as we will see later, it may well eventually collapse, due to regulation or governance problems resulting from the relegation of States, the regulators par excellence, to a secondary role in favor of large corporations. The DNA of these latter corporations is made up of nothing more than profit maximization and the profitability of their own activity. This is precisely, to our mind, the regime's distinctive feature.

   In addition to this essential trait, the new food regime has the following characteristics: the financialization of the economy (context), resulting from the slowdown of the real economy's growth; the rise of large transnational corporations, that have accumulated power in an unprecedented way, even coopting the power of states that legislate in their favor; in our case, global food corporations have emerged, turning into the main players of the food regime and international food markets. The establishment has adopted international agreements based on strict neoliberal regulations on a global scale, resulting from the hegemony of such corporations. They have thus been granted supremacy even over national legislations. The Uruguay Round and the creation of the WTO are the basic instruments of the creation of this international order. The result has been the increased industrialization

---

[3] These data (Schmidhuber, 2006) refer not only to the amount of ingested energy but to the total supply of dietary energy (dietary energy supply, DES, or gross diet). If we take into account MAPA data (www.mapa.gob.es/es/alimentacion/temas/consumo-y-comercializacion-y-distribucion-alimentaria/panel-de-consumo-alimentario/series-anuales/default.aspx) on the "quantification of the food diet in 1999" which only includes the food actually consumed, the figure would fall to 190 million GJ for the national total, on the basis of 2.768 kcal per person and year.

and specialization of global food chains, where national food systems are integrated into a new global labor distribution. The CFR is "pivoted on the internalization of neoliberal market principles by states subject to privatization via mandated structural adjustment and free trade agreements – as an alternative to a stable, hegemonic international currency" (McMichael, 2013, 15); it relies on the discourse of a neoliberal "globalization project", and is legitimized by arguments of market rule (novelty), agricultural "modernization" (a continuity), and the westernization of diets (McMichael, 2013, 47).

From the production perspective, the CFR is characterized by a form of accelerated technological innovation seeking to maximize large corporations' dominant positions (GMO seeds, for example) and to increase farmers' dependence. Agricultural activity thus turns into a source of permanent accumulation that maximizes profits at the expense of farmers' incomes. From the perspective of distribution, the regime is characterized by the so-called "supermarket revolution" (Reardon et al., 2003), the growing power of distribution companies that sell their products through big supermarkets and control a large part of the food chain. These large distribution companies have consolidated their power by selling their own brands or by diversifying their offer, meeting new demands of ready-to-eat meals, organic and certified food, and so on. From the consumption perspective, the regime has led to increasing food consumption differentiations between rich and poor consumers (Friedmann, 2005, 229), that is, the social segmentation of diet. Generally, only the highest-income social groups have access to the healthiest, more expensive food, i.e., food that is consumed fresh and of better quality, including organic food. For its part, lower-quality food, with high contents of fats, dairy products, meats, and processed products, is much cheaper and is consumed by low-income social groups. This explains their higher malnutrition incidence, causing an overweight condition and obesity among the working classes, with its associated diseases (diabetes, cardiovascular, etc.). The adoption of a diet rich in meat and dairy products, not only by rich countries but also by emerging economies such as China or India, is based on the so-called soybean–livestock complex (Berstein, 2010, 79). Raw materials (corn and soybean) within this complex originate in many countries in the Americas, allowing the market to expand due to the low price of grains in international markets. In this sense, the shift toward a meatier diet is not only a consequence of per-capita income growth in those countries, but also because of its very competitive end price thanks to the low cost of raw materials (González de Molina et al., 2017; Infante-Amate et al., 2018b).

3 A Regime on the Road to Collapse

It is ever more apparent that we are experiencing a unique crisis, one that is exposing the limits of modern civilization (Garrido Peña et al., 2007; Toledo, 2012a). There is an ever-more obvious contradiction between economic growth as a model of economic organization and the limitations imposed by the depletion of resources and the deterioration of environmental services. The scientific community warns that certain lines have been crossed regarding the ability to restore ecological dynamics on a planetary scale. The concept of "planetary boundaries" was developed by Rockström et al. (2009b) to assess whether we are in a "safe operating space" for humanity along nine key dimensions. Researchers suggest that four boundaries have already been exceeded (Steffen et al., 2015b). The 2008 crisis was, after the 1929 crash, the worst crisis of contemporary capitalism in both quantitative and qualitative terms. Unlike the first crisis, which was an overproduction and distribution crisis, the latter was an overconsumption crisis (increase in per-capita consumption of matter and energy) and put limits to the appropriation of natural resources. The triggers of the crisis not only included the Lehman Brothers bankruptcy and subprime mortgages but also the drastic price increase of raw materials, especially between 2000 and 2007. In just seven years, they grew by 277% (World Bank, 2009).

Thus, the crisis cannot pass for just one more of capitalism's cyclical crises. Its roots are structural and its effects are multidimensional. It has not only had an impact on the financial system, but on the very metabolism of the world economy, affecting both the extraction and consumption of materials. The mechanism that made it possible to control social entropy, that is, made social inequality more socially acceptable, is increasingly difficult to sustain. After the Second World War, the capitalist world order tried to recover economic growth by reducing the food, energy, and materials needed to fuel the reproduction of capital, that is, by intensifying domestic extraction not only in "developed" countries but also in peripheral countries, extending the frontier of appropriation of resources. Profits could thus grow again, warding off threats of social entropy by means of nominal wage increases and mass consumption. Postwar "social peace" achieved in industrialized countries came at the expense of a planetary increase in physical entropy. The exponential growth of energy and material consumption that led to the so-called Great Acceleration (Costanza et al., 2005; Hibbard et al., 2007) is a reflection of this. As Ulrich Beck (1998) vividly put it, social inequality did not disappear, but "climbed upstairs". Yet the current crisis is showing that it is increasingly difficult to reduce social entropy by increasing phys-ical entropy. The environmental crisis threatens to ruin the key mechanism that made capital accumulation and economic growth work, that is, the appropriation and cap-italization of nature (Moore, 2015).

A significant number of social scientists have read the crisis idealistically, attributing its origin to the "bad behavior" of political and economic elites or to financial regulation deficiencies. However, from a complex materialist per-spective (and that's what the ecological paradigm consists of), the crisis arises from the very core of capitalism. It is based on a rationale of constant profit rate increases and continuous growth of energy and material consumption that makes it possible. Both mechanisms, however, run into internal limits (the downward trend of profit rates) and external limits (the depletion and deterioration of nat-ural resources). Consequently, the crisis should be considered as a *metabolic crisis* (Garrido, 2015). Capitalism shows no signs of changing course, perhaps because this deeply engrained mechanism of interentropic compensation is what makes it work. Capitalism is promoting a way out of the crisis that consists of increasing the extraction of resources ever more, causing more environmental damages, i.e., stealing future generations and the world's poor of increasingly scarce resources, further accentuating territorial and social inequality (Piketty, 2014). If this course is not redressed, the hypothesis of a collapse of the civilizing process cannot be ruled out, at the cost of great social suffering. In the following sections, we describe the metabolic nature of the crisis and explain its causes. Our analysis later focuses on the corporate food regime (CFR), where contradictions are clearer still and the danger of collapse is nearer.

3.1 THE PHYSICAL IMPOSSIBILITY OF ECONOMIC GROWTH

The world population has doubled since the early 1970s, from 3,700 million to more than 7,600 million today. The growth rate has been 1.6% per year. The world economy's growth rate over the same period has been higher, rising from $15.7 trillion in 1970 to $52.9 trillion (at 2005 prices) in 2010. It has thus grown at a rate of 3.1% per year. In 2017, world GDP reached $80.6 billion at current prices (World Bank, 2019). However, since the beginning of the century, the price of many natural resources has begun to rise, creating a new economic context of higher and more volatile raw material prices, due to the scarcity or increasing costs to appropriate them. The era of cheap raw materials that sustained economic growth during the twentieth century seems to be over (UNEP, 2011; 2016, 29). Since the beginning of this century, the population and the world economy have grown at lower rates than during the second half of the last century, but material extraction has, on the contrary, accelerated. The relationship between both phenomena seems obvious and emphasizes the structural nature of the current crisis.

In fact, the annual use of materials on a global scale reached 70,100 million tons (70.1 Gt) in 2010, compared to 23.7 Gt in 1970. Contrary to population and world GDP, the use of materials accelerated between 2000 and 2010, reaching a growth rate of 3.7% despite the crisis. The extraction of fossil fuels grew by 2.9%, metallic minerals by 3.5%, and non-metallic minerals by 5.3%. Extraction of biomass is the only to have remained constant, at 2% (Krausmann et al., 2017a), highlighting the productive limitations of agroecosystems, as we will see later.

Globalization continues to progress, driven by the need to mobilize all natural resources. The metabolism of national economies is increasingly dependent on global flows of goods. In fact, international trade of materials has grown faster than world GDP, at an annual rate of 3.5%. A large part of these energy and material flows are invested in the creation and maintenance of stocks (buildings, infrastructure, equipment and machinery, etc.), especially in rich countries, further increasing the distance between some countries and others in terms of development and well-being.

These data disprove the existence of the famous "decoupling" (UNEP, 2011) of wealth creation regarding the consumption of materials. The decoupling gave rise to extensive academic and political literature, and even to the emergence of the so-called "green economy" that believed it was possible to maintain economic growth thanks to the decreasing consumption of materials. The drop in the energy intensity of the major economic processes, which had been occurring throughout the twentieth century, supported this assertion. However, the data on material extraction and consumption bring us back to the harsh reality, shattering the dream of indefinite economic growth. Per-capita consumption of materials on a global scale, i.e., the planet's metabolic profile, has grown strongly since 2000, despite the economic crisis. It has risen from 7.9 to 10.1 t in 2010 (Krausmann et al., 2017a). This growth has been faster than the growth of world GDP, showing that global efficiency in the use of materials has begun to decrease for the first time in a hundred years. Indeed, since 2000 there has been an increase in the intensity of the use of materials in the global economy: if in 2000, 1.2 kg of materials were required to produce a dollar, in 2010, almost 1.4 kg are required. Efficiency gains have been reversed during the period of great consumption acceleration since the beginning of this century. The phenomenon seems to be related to greater outsourcing of materials' extraction and processing activities to third countries, i.e., the "dirtiest" and most inefficient processes in terms of material consumption and waste production (UNEP, 2016; Krausmann et al., 2017a).

As we have seen, material consumption reached 70 Gt/year in 2010, that is 19 times more than in 1850, almost seven times more than in 1900 and five times more than in 1950. Biomass accounts for more than a quarter of extracted materials (27.15%), fossil fuels almost one-fifth (18.58%), and the remaining 54.29% are extractions of metallic minerals and especially non-metallic ones (Krausmann et al., 2017a, 656). The spectacular figure of mineral growth is due to rising needs of materials for building and maintaining "capital" stock, especially in rich countries. These stocks function as dissipative structures that provide services to society, with large amounts of energy and materials invested in their building and maintenance. A recent study estimated these needs at 26 Gt/year for 2010, the stock of the global economy being estimated at around 800 Gt (Krausmann et al., 2017b). These stocks, therefore, are responsible for the steady rise in energy and material demand. Any degrowth and sustainability strategy thus depends on the type of dissipative structures or stocks built and their dissipation needs.

As suggested, observable differences in different countries' "development" and welfare levels are related to the quantity and quality of the stocks that they have built and maintained over time. In fact, differences in per-capita consumption of energy and materials are linked to the demand generated by these stocks. This could explain the differences in "development" between rich and poor countries reflected in the differences in levels of per-capita consumption. For example, consumption in industrial countries or rich countries is on average 15.9 t/capita/year, while in peripheral or "less-developed" countries, average consumption is 2.6 t/capita/year (Krausmann, 2017a, 657). However, these material domestic consumption figures must include the cost of extraction and initial processing of materials that are eventually exported to industrialized countries. In fact, in many net exporting countries of energy or materials, a portion of the extracted materials is invested in the extraction process itself and not in the endowment of stocks for the benefit of their own population. That portion of energy and materials can no longer be used to build and maintain stocks in the future.

Indeed, economic growth experimented by developed countries is maintained thanks to appropriation and consumption of low-cost natural resources from less-developed countries, and the consequent deterioration of their standards of living and environmental quality conditions. This pattern rising from unequal economic development and aggravated by socioecological effects of international trade, resembles the old dichotomy between development and underdevelopment of the 1960s' and 1970s' economics literature, today rescued by the metabolic analysis of world economy. Many top consumers of resources have a lower domestic extraction than their total consumption of energy and materials. The development of transport and colonial policies made industrialized countries growingly dependent on resources from peripheral territories, to the point that their economies began to increasingly depend on external inputs (Hornborg, 2007; Martínez-Alier, 2007). With globalization, international trade has become a crucial instrument for almost every country in the world. The larger a country's metabolism, the greater its dependency on international markets for provision of natural resources not available from their own domestic extraction. For decades, sources of energy and materials – generally located in less-developed and peripheral countries, but not necessarily – have been essential for sustaining the metabolic process of developed economies.

Schaffartzik et al. (2014) have shown that global exports increased from 0.9 Gt/year in 1950 to 10.6 Gt/year in 2010. A total of 40% of fossil energy carriers, 35% of ores, 8% of biomass, and 3% of non-metallic minerals are transferred via international trade across countries (Krausmann et al., 2017a, 657). As expected, Europe, North America, and more recently also East Asia, are the biggest net-importing regions, whereas all other world regions are net-exporting regions (UNEP, 2016). China is the country where the consumption of materials has grown fastest, going from 6 tons/capita/year in 1995 to 17 t/capita/year in 2010. Exports rose from 7% of global domestic extraction in 1950 to 16% in 2010. What international trade actually does is displace environmental burdens from one country to another. In accordance with these trends, the global biomass market is dominated by the so-called "flex crops": wheat, corn, and soybeans. The market situation, quality and price determine whether these commodities are sold as foodstuffs, biofuels, or animal feed. The biggest global commodities of this type to come next are sugar, palm oil, and rice, reflecting, with the exception of rice, the growing needs of food-processing industries.

The general trend is thus to use imports to compensate growing demand that cannot be supplied by domestic extraction (Bruckner et al., 2012). Resource use is distinctly unequal. Because of this, some academics have denominated this phenomenon as "unequal ecological exchange" (Hornborg, 2011). Poor countries export resources that have a low value but that represent a large volume in the world's economy. This latter conclusion was reached by studying the physical trade balance (PTB) of several developed and underdeveloped countries (Fischer-Kowalski & Amann, 2001; Muradian & Martínez Alier, 2001; Giljum & Eisenmenger, 2003; Muñoz et al., 2009). A recent analysis was made of the PTB of most world countries between 1962 and 2005 (Dittrich & Bringezu, 2010; Dittrich et al. 2011), and the conclusions agreed with findings of previous works: since the 1960s, more industrialized regions had positive PTB values, while developing regions had negative PTB values. Such patterns gradually intensified until 2005 (Dittrich & Bringezu, 2010). In other words, resources continue to flow from south to north. According to Krausmann et al. (2017a, 657), the shift from industry to services in industrial countries, high labor costs, and rigorous environmental standards result in externalizing resource-intensive industries to the Global South where raw materials are processed and "light" products are exported to meet the needs of high-tech industries and consumers in the industrial world.

3.2 METABOLIC CRISIS AND CAPITAL ACCUMULATION: A MARXIST READING

The effects of natural resource depletion and the worsening of the environment that make economic activity possible go beyond the health of ecosystems. There are direct consequences on the relative prices of food, raw materials, and energy, as their prices rise. As Jason Moore (2015) argued, the era of cheap nature has come to an end, and this directly affects the very heart of economic growth and capital accumulation. The extended reproduction of capital faces difficulties in a context of increasing scarcity of energy and materials and of increasingly adverse environmental conditions. The industrial metabolic regime itself is put into question and with it, the social system that implanted it: capitalism. Marx ventured an economic explanation of its contradictions that can also serve to reaffirm the structural nature of the metabolic crisis.

The tendency of the rate of profit to fall (TRPF) is possibly Marx's great scientific contribution. Paradoxically, it is also Marx's most questioned thesis, even by a major part of later Marxist discourses, despite the fact that TRPF constitutes an empirically verifiable normative formulation. Although the idea had already been advanced by David Ricardo, Marx made it theoretically coherent within the framework of his general theory of capital. It is well known that profit is the result of dividing the rate of exploitation by the organic composition of capital. Because an organic composition greater than capital is needed to obtain equivalent exploitation rates, the profit rate tends to decrease. This trend can be interpreted in biophysical terms, given that the organic composition of capital is directly related to the extraction of natural resources, their availability and their price. The downward trend of the profit rate drives the search for new frontiers of appropriation of natural resources, encouraging their cheapening. However, the horizon of depletion and the steady loss of extraction process efficiency make resources increasingly expensive and, therefore, lead to a greater organic composition. For example, it is increasingly expensive to extract the labor equivalent of one ton of oil (Hall et al., 2009; Hall, 2011). Capital, like a drug addict, needs to consume ever-higher doses to obtain equivalent pleasure, which fatally also decreases. Capitalism's addiction to growth is motivated by the implacable logic of the marginal tendency of profit to fall (Brenner, 2009; Tapia & Astarita, 2011).

In turn, the pressures to lower the cost of the organic composition of capital have unfolded on two fronts. The first front is technological, with the efficiency increase of the technologies used (fixed capital). The second front is political-institutional (military, in some cases) and covers the control of raw materials prices, the drastic reduction of fiscal pressure on large fortunes and extensive deregulation. However, the fixed capital reduction deriving from technological improvement has led to rising production because of lower costs, reducing overall efficiency as predicted by the "Jevons Paradox" (Alcott, 2005). This has resulted in expanding global consumption of materials and energy due both to raw materials consumption growth by emerging powers – through increasing production and internal consumption – and to the outsourcing, by most industrialized countries, of processes that use materials more intensively, as described earlier. All this has reinforced the prospects of scarcity and consequently increased raw materials' intertemporal discount rates due to their foreseeable depletion. These discount rates have in turn affected raw material prices, raising the costs of the organic composition of capital.

In this context, the TRPF's avoidance strategy has aimed at increasing the exploitation rate through an increase of absolute surplus value. Offshoring and globalization have transitorily managed to reduce wage costs using quasi-slave labor in developing countries and lessen the weight of nominal wages in the developed West. Reducing Western wages would have had to carry an undesirable consequence: the fall of domestic consumption. This consequence, in addition to being economically undesirable, would have been politically unbearable in democratic regimes. To counteract these perverse effects, real wages have been augmented in a fictitious way by means of three factors: the invasion of delocalized products from emerging powers (China, India, Brazil), lenience toward, and even stimulation of, speculative bubbles such as the real estate bubble (with its resulting wealth effect), and the irrational increase of credit and indebtedness. These instruments are unlikely to be maintained for a long time. The productivity growth of emerging powers has led wages and domestic consumption to increase (Garrido, 1996); this has not only countered the cheap product supply, it has also increased the internal consumption of energy and materials, aggravating the prospect of scarcity. The other two instruments broke up with the financial crisis unleashed at the end of 2007.

Okishio's theorem (1961) has represented the most accurate objection to the TRPF. The theorem states that the profit rate (referring actually to the profit margin) increases with technological innovation, because the costs of fixed capital go down, and this drop at constant wage costs would bring about a decrease in the value of the organic composition of the capital and an increase in relative surplus value. In this way, investment in technological innovation would entail a comparative advantage that would thus be enhanced. If we observe how the weight of wages has dropped in relation to GDP and the negative deleveraging of wages and labor productivity (increase in negative surplus value), the theorem would be right and the TRPF would be wrong. But Okishio's theorem would not be correct if we take into account two key factors: the impact that wage containment (and the increase in relative surplus value) has on consumption and on the increase of social inequality, and the rising costs caused by the raw materials' rise in cost due to their depletion or inefficient extraction rate in a context of growing demand. That is, Okishio's theorem would be correct if consumers and natural resources were infinite, but they are not. The new theoretical model of the TRPF, the so-called "temporal single-system interpretation" (Laibman, 1992; Tapia & Astarita, 2011), supports the empirical validity of the falling trend against the objections of Okishio's theorem by introducing time into the calculation of the tendency for the profit rate to fall (Hendrich, 2017).

As a result, it is becoming ever-more difficult to increase profit rates because the availability of resources and the quality of environmental services are raising the costs of the organic composition of capital. The internal devaluation undergone by many industrialized countries, due to deteriorating wages and labor rights, could be largely understood as a response to this trend, expressed in terms of competitiveness in global markets. In other words, it is increasingly difficult to compensate the social entropy generated by the capitalist system using the consumption of energy and raw materials, that is, using an increase in physical entropy. Future scenarios are not very promising. The system has reached extraction and consumption levels that are hard to sustain. It has substantially reduced the chances of completing a planetary transition to an industrial metabolic regime. In other words, the metabolic regime can only postpone the collapse at the expense of deepening social and territorial inequality, that is, increasing social suffering (social entropy).

Indeed, the discourse of international organizations and of most governments maintains that economic growth is the only way in which developing countries can progress and achieve the same levels of well-being that industrial countries have already achieved. They argue that there no employment or wealth distribution is possible without growth. Some politicians and intellectuals from these poor countries claim their right to enjoy a standard of living similar to that achieved in industrialized countries. It is a matter of equity. However, although access to the same levels of prosperity as in the West is a reasonable and socially just goal, it is unviable from a biophysical perspective. It is not possible to generalize the consumption levels of industrialized countries, nor can this be achieved through economic growth. The data make that clear. Were the entire planet to reach Europe's current average metabolic rate, approximately 16 t/capita/year around the year 2050 – when the population will reach over nine billion people – the world's consumption of materials would reach 140 Gt per year, almost three times more than today (Krausmann et al., 2017a, 663). Some studies raise that amount to 180 Gt/year or 20 t/capita/year by 2050 (Schandl et al., 2016). This would mean that many countries would have to increase their current metabolism fivefold. It would be necessary to double the use of biomass, quadruple the use of fossil fuels, and triple the annual use of minerals and construction materials. Carbon per-capita emissions could triple and global emissions could quadruple to 28.8 GtC/year, surpassing the most pessimistic scenario of emissions calculated by the IPCC (Krausmann et al., 2017a, 663). It is neither physically nor socially possible to complete the socioecological transition to the industrial metabolic regime: nature simply does not allow it.

In view of the depletion of resources and the worsening of environmental services, the logical response would be to reduce the metabolic profile of industrialized countries and distribute the consumption of energy and materials more equitably. As shown by Bringezu (2015), "business as usual" cannot be considered a sustainable level of global material use. The model's non-viability is not under discussion; what *is* under discussion is for how much time it can be maintained without major structural reforms that will transform it from top to bottom. The ongoing rise of CO_2 emissions or the extraction and consumption of materials speak sufficiently of their own unfeasibility, not to mention regressive environmental policies adopted by the Trump or Bolsonaro governments, for example. The planet faces a serious problem of governance. Private interest has become the supreme interest that governs humanity's destiny. National governments are in the hands of the corporations themselves through what Barrington Moore (1966) called "service classes". No institutions democratically govern global socioecological problems and propose solutions to reverse the crisis.

As a result, the likelihood that things will remain the same is very high, possibly leading to a collapse of the civilizing process and a significant rise in social suffering. There are no guarantees that the market itself, in its current phase of wild deregulation, can avoid it. The most probable route is that this state of affairs will be prolonged, further deepening social inequality and ecological deterioration. Indeed, it is very likely that metabolic entropy will continue to increase (economic growth) to the benefit of industrialized countries to partially compensate for social entropy (maintaining and even increasing high levels of domestic consumption) at the expense of the natural resources of peripheral countries, that is, of poor countries, whose social entropy will continue to increase, naturally in the knowledge that this solution has ever-more obvious limits. The migratory phenomenon, now global, is but the visible response of poor countries to the plundering of their resources and the socioecological impact of extractivism. In other words, without a major change in the metabolic patterns of energy and materials extraction and consumption, the possibility of collapse will no longer be a remote possibility or a mere working hypothesis.

3.3 INDUSTRIAL AGRICULTURE: AN INEFFICIENT AND HARMFUL MODEL THAT IS EXHAUSTED

The industrial agriculture crisis shares the same causes as the global crisis, but specific factors make it even deeper and the risk of collapse greater. Economic policies aimed at lowering food prices in order to lower wages and promote economic growth in other sectors of activity are responsible for the downward trend of prices received by farmers. There have been attempts to compensate this trend with successive production increases based on a greater use of inputs, i.e., an increase in the organic composition of agricultural capital. The rising cost of raw materials and fossil fuels with which agricultural inputs are made has further depressed farmers' incomes and turned this compensation mechanism into a vicious circle in which income has tended to decrease, at least since the Second World War. One could say that the agrarian sector has been parasitized due to the transfer of surplus value to other economic sectors via product and input markets, meaning that family farmers, based on the income they have received, have reproduced their household economies with difficulty. We illustrated this convincingly in the case of Spain (González de Molina et al., 2019). The organic composition of capital has effectively grown in the agrarian sector as a way to raise exploitation rates (profits), not on the basis of relative surplus value, however, but on the basis of low income or remuneration of work, i.e., based on absolute surplus value. This characteristic has made industrial agriculture even more vulnerable than other economic sectors.

In effect, the human right to food (FAO: www.fao.org/right-to-food) is compromised by the pretensions of the CFR (McMichael, 2006), which is more concerned with generating benefits than guaranteeing that right. Hunger has not disappeared. There are still 821 million undernourished people around the world (FAO, 2018). A large part of the problem is caused by uneven distribution of food. Half of those who go hungry are peasants and farm workers. Low prices and lack of access to land and other natural resources are the main factors explaining hunger and rural poverty. Industrial agriculture is making them even poorer by depriving them of markets, expropriating their land and water, and polluting their soil (Heinrich Böll Foundation, 2017). Meanwhile, the world harvest of edible crops provides around 4,600 kcal per person per day, but only around 2,000 kcal per person are actually available for consumption. A significant part goes to feed livestock and biofuels and the rest is lost along the food chain (Lundqvist et al., 2008).

Moreover, the agrifood sector generates perhaps the biggest ecological footprint in the world, acting as the main driving force of the planet's biophysical transformations (Tilman, 2001; Foley, 2005; Weis, 2013; Steffen et al., 2015b). Global assessments of the current agriculture and food system suggest that the system is a major driver of the process leading us to the planetary boundaries for climate change, biogeochemical flows, and biodiversity loss (Campbell et al., 2014). The contradiction manifests itself in the growing difficulties to continue expanding the volume of food production in increasingly degraded agroecosystems and the use of ever scarcer resources and services (oil, phosphorus, climatic stability, etc.). Despite this, the pressures to increase production are still fueled by institutional structures and by wealth distribution, threatening to lead to the general failure of the food regime as a whole.

Indeed, four factors make the crisis of the industrial agrarian production model, underlying the CFR, a structural one: (i) the slowdown in agricultural growth; (ii) the increasing lack of profitability of agricultural activity, a result of the industrial model's very configuration; (iii) the fact that the model is based on the use of agrochemicals, machines and water elevation and conduction systems, making it highly dependent on fossil fuels, which in turn are increasingly expensive and scarce; and finally (iv) despite the fact that the degree of artificialization of agroecosystems continues to grow, agricultural activity is very dependent on environmental conditions and therefore very sensitive to climatic events, which are becoming more frequent and extreme due to ongoing climate change. The first two factors are internal to the model itself, while the latter two are external and depend on general economic progress, both on the consumption of fossil fuels and the emissions they generate. To the extent that the latter two factors are beyond the control of the agricultural sector, we will focus on the internal factors, which are the most interesting from an agroecological perspective.

The extension of worldwide cultivated land increased by a third during the twentieth century according to Smil (2001, 256), while productivity increased fourfold, thus allowing a sixfold growth of harvests in the same period. However, we have been witnessing a slackening of food production growth in recent years. Between 1950 and 1990, production per hectare grew at a yearly rate of 2.1%, while it grew by only 1.3% between 1992 and 2005 (FAO, 2007). In fact, biomass extraction has grown at a constant rate of 2% per year since the 1970s, and, in comparative terms, biomass is the material that grew the less, as described earlier. The figures reflect the limitations encountered by the CFR to keep growing at the pace of other economic sectors. Despite this, biomass extraction today entails the appropriation by the human species of approximately 25% of the planet's annual net primary productivity (Krausmann et al., 2017a, 649).

Some types of wood and firewood are the types of biomass that grew the least since the 1970s, that is by 76% and 32%, respectively, due to the appearance of cheaper alternatives to wood as a material, the reduction of paper consumption, and the substitution of firewood with other types of fuel. In contrast, biomass from pastures grew faster, a behavior also proper to fodder crops. If taken together, both categories grew by 131%, due to an increase in the consumption of animal products. Domestic extraction of crops for the production of sugar (sugar crops) showed higher growth percentages (137%), resulting from the growing importance of processed food and what the statistics call "other" biomass, which groups vegetables and oil crops, and grew by 150%. This group has greatly expanded over the last 40 years.

The international biomass trade has followed the same globalization pattern as the rest of the materials. The biomass circulating in international markets has risen from 370 million tons in 1970 to 1.9 billion tons in 2010, growing at a rate of 4.2% per year, although the rate has fallen to 3.2% since 2000 (UNEP, 2016). Global trade in agricultural products grew sixfold from 1960, while production only grew twofold (Mayer et al., 2015). On the whole, biomass trade has grown at a higher rate than average international materials trade, showing that biomass flows have undergone a very intense globalization process, encouraged, we believe, by the growing limitations that agroecosystems pose to CFR plans.

In fact, at least four factors underlie the growth slowdown. If the current state of affairs continues, these factors threaten to limit world food production even more in a context of increasing population and consumption. In the first place, the possibilities of adding more agricultural land seem to have been reduced. As shown in Figure 3.1 and Table 3.1, the global surface area devoted to agriculture grew strongly until the mid-1990s and reached its peak around the year 2000. Since that year, agricultural land has been reduced by 81.4 Mha (1.63%) and surface area devoted to livestock feeding by 140.2 Mha (4.1%). Although new agricultural land forecasts are optimistic (they would increase by 5% until 2050, according to the FAO), this does not seem to be a viable means to substantially increase the volume of world production (FAO-HLEF, 2009, 9). The stock of land useful for agricultural activity as a whole is limited and cannot grow indefinitely. Furthermore, most of it is located in Latin America and sub-Saharan Africa, where lack of access and infrastructure

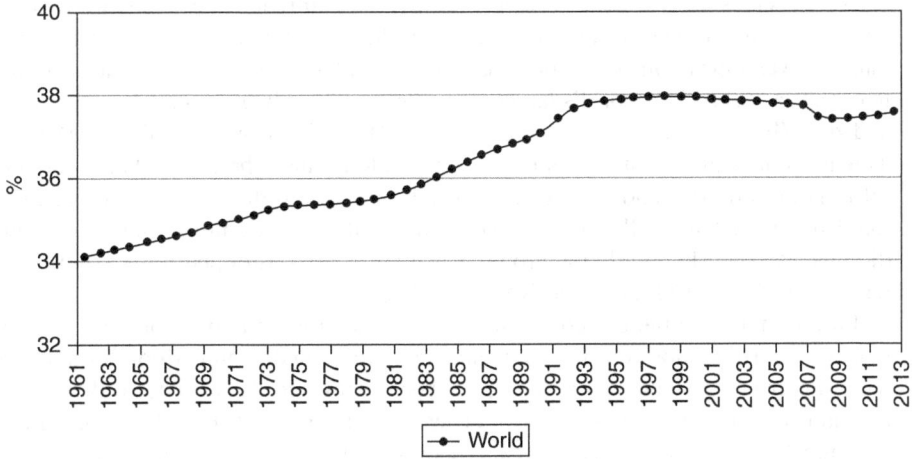

FIGURE 3.1 Agricultural land share of land surface.

TABLE 3.1
Global Evolution of Main Land Uses (Million Hectares)

Year	Agriculture Mha	%	1961 = 100	Pasture land and prairies Mha	%	1961 = 100	Irrigated land Mha	%	1961 = 100
1961	4,457	34.1	100	3,077	23.6	100	161	1.2	100
1970	4,565	35.0	102	3,128	24.0	101	184	1.4	114
1980	4,649	35.6	104	3,197	24.5	103	221	1.7	137
1990	4,831	37.0	108	3,302	25.3	107	258	2.0	160
2000	4,954	38.1	111	3,417	26.3	111	287	2.2	178
2010	4,868	37.4	109	3,321	25.5	107	321	2.5	199
2016	4,873	37.5	109	3,276	25.2	106	334	2.6	207

Source: FAOSTAT (accessed November 6, 2018).

could limit their use, at least in the short term. It would be fair to believe that given the limited amount of land availability, productivity may still rise by means of a new expansion or consolidation of irrigation water. The irrigated area has doubled since 1961, from 161 Mha to 334 Mha in 2016. Irrigated agriculture covers one-fifth of arable land and contributes nearly 50% of crop production. However, the interannual growth rate has been slowing down since the 1990s. The increase in irrigated land has fallen way below the population growth rate, to such an extent that irrigated land per-capita shifted from 0.052 hectares in 1961 to 0.045 hectares in 2016 (FAOSTAT, accessed November 6, 2018). This option may also be hindered due to growing water shortages prevalent in many areas parts of the world and 1.4 billion people live in areas with sinking ground water levels (FAO-HLEF, 2009, 9–10).

If we set aside the disruption global warming might be causing to weather conditions,[1] it can be said that the world's production of cereals depends largely on land and water availability. It seems that prospects of an increase of the land allocated to cereals are dim. The world's land sown with cereals only grew 11% between 1961 and 2016 (FAOSTAT, accessed November 6, 2018), going from 647 Mha to 718 Mha. This is a consequence of soya's new leading role to meet the demands for cooking oil in poor countries and animal feed in rich ones. According to the forecasts, this trend in cereal land will continue. The availability of agricultural land per capita diminishes as world population grows. Since 1961, it has dropped from 1.44 to 0.65 ha in 2016 (FAOSTAT, accessed November 6, 2018).

Furthermore, the intense competition established by urban development projects that cover fertile lands contributes to enhancing this trend. Competition for land use has been growing since energy uses have been recovered. It is expected that global demand for meat will grow by 50% at least until 2030 (FAO, 2008), thus adding pressure to the production of cereals (see below). In 2010, a total of 14 million hectares, about 1% of agricultural land, were allocated to the production of agrofuels. It is predicted that this amount will reach 35 million hectares in 2030 as well, but the key factor that will largely determine the future scarcity or abundance of agricultural land and its production is not located on the supply side, but on the demand side. Ever more meat and dairy products are eaten, forcing the number of herds of cattle to rise to levels unseen before. Global meat production has undergone a dramatic increase, from 92 million tons in 1967 to more than 330 million tons in 2007 (IAASTD, 2009). These trends in food consumption are shared by most developed countries, even by emergent countries such as China or India. For their upkeep, plots of land formerly allocated to provide human food are used to grow feed and fodder for cattle fattening. The global appropriation of land biomass reached 18,700 million tons of dry matter per year, that is 16% of the land's net primary production according to Krausmann et al. (2008a, 471). Of this amount, barely 12% of the vegetable biomass went directly to human consumption, 58% was used to feed cattle, 20% was used as raw material in industry, and the remaining 10% as fuel. The combined action of these demands over a limited amount of land threatens to further increase the pressure on the Earth's ecosystems.

To all this must be added the environmental damage caused by agricultural activity on the productive capacity of the planet's agroecosystems. The degradation is turning vast cropland areas into soils unfit for cropping. The desire to produce large amounts of food, water, wood, fibers, fuel, and other materials is causing a profound impact on agroecosystems, and in all sorts of ecosystems on Earth. Natural habitats are being

---

[1] Both agriculture and food safety will be affected by global warming. It is predicted that in tropical and subtropical regions the impacts will result in a fall of the potential yields of crops (Nelson et al., 2009, 17). It is also foreseen that the availability of water resources will also decrease, in line with a higher widespread risk of floods caused by the rising levels of the oceans and more intense rainfalls. Other predictions include a dramatic increase in the frequency of catastrophic weather phenomena, such as hurricanes, typhoons, and droughts triggered mainly by big climate changes (Dixon et al., 2001, 5). In Asia, scientists studying plant behaviour very closely have concluded that the rise in temperature in the next 50 years may lead to cereal production reductions in the Tropics of up to 30% (Nieremberg & Halweil, 2005, 127).

transformed owing to the expansion of specialized agricultural, cattle, and forest systems. At present, half of the world's area free from ice has been turned into land suitable for crops, livestock, and forest production. Between 1700 and 1990, available agricultural land multiplied fivefold and the land allocated to livestock increased six times (Hibbard et al., 2007). The former process is to blame for the widespread deforestation that is taking place in tropical regions, including rainforests and mangrove swamps. On top of this, aquifers are being overexploited to cater for cropland, livestock, and forest produce. Moreover, fish populations and other sea organisms are also being overexploited by a predatory and unsustainable fish industry. As a result, sea populations are being depleted or neighboring this point in most oceans (FAO, 2000).

High-inputs intensification and productive specialization have triggered an intensive use of natural resources (land, water, biodiversity, etc.). Erosion, mineralization, loss of soil nutrients, deforestation, overgrazing, and inadequate agricultural practices are to be held responsible for the degradation of many croplands. Between the mid-1940s and 1990s, 1,970 million hectares, 15% of the world's surface excluding Greenland and the Antarctica, had undergone degradation according to calculations carried out by regional experts and published by the Global Land Assessment of Degradation (GLASOD, 1991; UNEP, 1991). More than 20% of agricultural land worldwide is now classified as degraded, with degradation progressing at an alarming rate of 12 million hectares a year, equivalent to the total agricultural land of the Philippines (Heinrich Böll Foundation, 2017, 33). Water resources are being severely affected by this intensive use as well. Agricultural production has grown over the last 50 years owing largely to water being diverted for irrigation, to such an extent that 70% of freshwater obtained from surface or underground sources goes yearly to cropland (WRI, 2002, 66). As we have seen, the world's irrigated area increased from 94 million hectares in 1950 to 334 million hectares in 2016 (FAOSTAT, accessed November 6, 2018). This massive use of water has implied the execution of large-scale works to divert, channel, store, and regulate surface waters together with the drawing of large quantities from underground aquifers. Besides the damage water channeling is causing, some of them at the root of many current natural disasters and ecosystems degradation, agriculture has led freshwater availability for its own use and for human consumption to fall sharply (UNEP, 1994). The UNEP (1994) estimated that in that year, 40 million hectares of irrigated land had been damaged by salinization, rendering recovery difficult and expensive for agriculture. Six years later, 100 million hectares were degraded by salinization, sodiumization, and hydromorphism. This situation, which is bad at present and will get worse in the future, is ringing alarm bells because the capacity of agroecosystems to produce food, raw materials, and provide environmental services is shrinking. For example, it is estimated that farmers incur losses of $11 billion a year on account of the damage salinization causes to the productivity of the land (WRI, 1999, 92). Based on the data gathered from GLASOD above, it has been calculated that the total losses in yield accumulated over the last 50 years as a result of soil degradation were 13% of total value in agriculture and 4% in grazing (WRI, 2002, 64). Most recently, Costanza et al. (2014) have calculated that the global change in land use between 1997 and 2011 has led to a loss of environmental services. They have valued this loss at between US$4.3 trillion and US$20.2 trillion per year, and believe that this estimate is quite conservative.

3.4   EVIDENCES OF AN ANNOUNCED COLLAPSE

The CFR thus shapes a way of managing agroecosystems that leads to collapse. It pushes farmers into a spiral in which a lack of agricultural profitability is compensated by the increase of production and production specialization; farmers use ever-more expensive external inputs to make this possible, further reducing profitability and favoring a form of new production intensification that eventually undermines the basis of the resources themselves. Industrial agriculture is therefore unable to provide sufficient income to farmers, except in the case of a few types of farms or regions of the world where it has done so partially. Agriculture's constant loss of profitability, owing both to the modus operandi of national and international markets and the sub-ordinate role that agriculture plays in the growing economies, is putting agricultural production at stake. In rich countries, public subsidies are used to partly compensate for this loss. The total value of the world's food, feed, fodder, and fiber production reached $1.5 billion in 2007 (FAO, 2007). Such an amount suggests that the food per person has increased by 16% since 1983. However, the same cannot be said of the incomes farmers obtain from agricultural prices, which have plummeted by 50% in the same period (FAO, 2007).

An increasingly slimmer percentage of the final price of agricultural products is going into the hands of farmers, increasing external inequality. The reasons are manifold, but definitely include a growing corporate concentration of market power in the distribution sector, together with the increasing role of processes of transform-ation, delivery, and processing. These activities require labor and capital, consume materials and energy, and generate wastes, thus enlarging the fraction of the value added in the final price of the product that remains beyond the reach of its produ-cers – as conventional economists say. The loss in profitability of internal agricultural production reflects the constant deterioration in external terms that trade agricultural commodities have been experiencing worldwide. As we have seen, this deterioration in agrarian profitability has been favored by the economic policies implemented by States practically all over the world in order to provide the rest of the economy with cheap food, while lowering the cost of wage labor. This double-sided process is responsible for the abandonment of agricultural activity in developed countries, and for poverty and the high migration flows to the cities in developing countries. It also responsible for the decline in agricultural employment: according to FAOSTAT, it dropped from 38% of employed population in the year 2000 to 30.7% in 2014.

As we said before, the decline in profitability is aggravated by uneven land dis-tribution and restricted access to land. Not all countries have agrarian censuses and those that do exist are often conducted following different methodologies. No con-sistent data thus exist on the structure of land tenure, but we do dispose of fairly reliable estimates that offer, nonetheless, a global and realistic picture. Lowder et al. (2016) conducted an exhaustive review of the available data on the size of farms and their distribution according to surface area on a worldwide and nationwide scale, as well as of the studies based on national data. Their estimate is based on the available data of 167 countries representing 96% of the world's population, 97% of the population active in agriculture, and 90% of agricultural land worldwide. The results of the study show that there are more than 570 million farms worldwide, most of which are small and family-operated. About 12% of global agricultural land are small farms (less than 2 ha) and about 75% family farms. More than 90% of the world's farms can be considered family farms, but 84% of all farms are small farms (less than 2 ha). According to this estimation, family farms (not the small farms alone), operating 75% of the world's agricultural land, are likely responsible for the majority of the world's food and agricultural production.

Another estimation, following a different methodology and a more conservative approach to the definition of family farms, is based on a census from 105 countries encompassing 85% of the world's food production. Results were different: family farms constitute over 98% of all farms, and operate on 53% of agricultural land, producing 53% of the world's food (Graeub et al., 2016, 1). In any case, both estimates show that the vast majority of holdings in the world are small and family farms.

Whichever estimate is taken into account, they both show that the structure of the holdings is highly unequal. There are two large groups: a small number of large farmers own a large amount of agricultural land (2% of all farms and 47% of agricultural land, according to Graeub, 2016) and are responsible for a large part of the biomass flows that sustain the CFR; the second group is the great mass of peasants and family farmers who work on small farms and who dedicate a major part of their products to maintaining their families and feeding their respective countries. The data also show that there are differences depending on whether the countries are rich or poor, are located in industrialized centers or in the periphery. Indeed, the regional distribution is also similar in both estimates, Asia being the continent with the largest number of holdings (74%); the majority of farms are found in lower- or upper-middle-income countries (representing, respectively, 36% and 47% of the 570 million farms worldwide). Thirteen per cent of farms are in low-income countries. Farms in high-income countries represent 4% of the world's farms. In countries with lower income levels, smaller farms operate a far greater share of farmland than do smaller farms in the higher-income countries (Lowder, 2016, 27). According to data from the last Brazilian Agricultural Census, carried out in 2017, a total of 50,865 large holdings hold 47.6% of the agricultural land, with an average area of 3,300 ha. This means that around 1% of holding companies own almost 50% of the total agricultural area. On the opposite side, small holdings (below 10 ha) represent 50% of the total but cover only 2% of the land, with an average area of 3.14 ha (IBGE, 2018). According to a report by CEPAL, FAO, IICA (2012), census data from Chile, Argentina and Uruguay – where the agricultural sector is clearly oriented toward commodity exports – show a tendency toward land concentration. Something similar can be said of other Latin American countries (FAO, 2012a).

Lowder et al. (2016, 27) also calculated average holding size and reached the conclusion that average farm size decreased in most low- and lower-middle income countries from 1960 to 2000, whereas average farm sizes increased from 1960 to 2000 in some upper-middle-income countries and in nearly all high-income countries for which we have information. The result for the low- and lower-middle-income countries mentioned is due to population growth, reduced access to land, and rural poverty; in the case of upper- and middle-income countries, it owes to lack of farm profitability that pushes toward abandonment and the increase of the average size of holdings to reach a minimum profitability threshold. This mixed trend is forecast to continue in the future (Lowder et al., 2016, 27). The farm structure has become more unequal since 1960 and it seems this inequality is set to rise even further in the future.

The rise of inequality will have consequences regarding sustainability. As we saw in the previous chapter, inequality of land distribution and income has been responsible for the process of intensification and productive specialization that has led to the current crisis. Consequently, the tendency revealed by the data is that the pressure on agroecosystems to produce more or to specialize in products with market outlets, accentuating the tendency toward monoculture, will be reinforced together with the degradation of agroecosystems. If, as it seems to be the case, the regime continues to maintain the downward trend of prices perceived by farmers and peasants – given that this is a structural characteristic, as described earlier, of the capitalist regime – it seems clear that activity abandonment will continue to take place in the north and rural poverty will continue to grow in the south, where the agrarian sector's economic dependence is highest and the rural population has the greatest weight.

The use of Green Revolution (GR) technologies is no longer an option either to increase production or to increase income. The latter depends on an institutional arrangement that will not be easily changed, while production increase using more chemical fertilizers, pesticides, and fuel consumption is uncertain and only deepens the downward spiral of profitability with ever smaller yields. In recent years, yield increases for key crops have started to plateau in various regions of the world, e.g., in the US or Japan. A meta-analysis of yield developments around the world from 1961 to 2008 found that in around one-third of the areas that grew maize, rice, wheat, and soybeans, yields either failed to improve, stagnated after initial gains, or even fell (Heinrich Böll Foundation, 2017, 33).

GR technologies are no longer capable of achieving a substantial increase in the volume of agricultural production as a whole, although the use of fertilizers has multiplied sixfold since 1961, totaling sales of $175 billion and some 246 million tons in 2013. It is expected that the amount of fertilizers will reach 273 million tons in 2020. Most of these amounts consist of nitrogenous fertilizers (71.5%) originating the widespread nitrate contamination of both surface and underground water courses. However, this growth is expected to occur especially in Africa, where it is more uncertain due to farmers' poor purchasing capacity, followed by Latin America and Eastern Europe. Conversely, in countries with greater purchasing capacity, their use shows obvious signs of stagnation, both because the marginal utility of a greater use is more than debatable and because of the legislation attempting to mitigate its adverse impacts. It is also due to the fact that developed countries, such as members of the EU, have reduced fertilized surface areas, significantly outsourcing the production of biomass, especially soybeans and corn, for animal feed (Witzke & Noleppa, 2010; Infante et al., 2018b). The fertilizer market of countries and regions such as China, North America, Western Europe, and Australia, which account for more than half of the global market, are expected to be saturated by 2021. According to the data provided by the Heinrich Böll Foundation (2017, 18), multinational agricultural trading groups such as Archer Daniels Midland, Bunge,

Cargill, and Louis Dreyfus Company have reduced their investments on the account of low growth prospects.

Pesticide use shows a similar behavior. More than 550 insect species have developed pesticide resistance. Over the last 50 years, an average of 13 new cases of resistance to these substances have been reported annually. Something similar has happened with weeds, which have developed resistance to each type of herbicide in use. Resistance is due to genetic and hereditary mutations, which in turn result from the selection process caused by the repeated use of pesticides. Resistant individuals survive and reproduce, so the percentage of survivors increases over successive treatments to the point where the pesticide loses its effectiveness (FAO, 2012b). Therefore, higher doses must be used to successfully treat both pests and diseases, and to control weeds. The vicious circle of increasing pesticide use and increasing resistance brings about mounting costs for farmers, as well as further environmental damage. However, current evidence suggests that the evolution of insects and weeds may exceed the ability to replace obsolete chemicals and other effective control mechanisms (Gould et al., 2018).

In short, the use of GR technologies is showing signs of depletion in its capacity to produce significant yield and production increases. This explains why the sale of these types of inputs has decreased or stagnated in countries with an industrialized agriculture. As we will see later, it is uncertain that the new generation of agricultural technologies will substantially increase yields per unit of surface area; furthermore, they also raise intermediate production costs, deepening the spiral of mechanisms generated by the crisis. Depending on expensive external inputs cannot be a solution.

The difficulties of agriculture have oriented food corporations, the production of inputs, and some farmers toward intensive livestock and participation in the "Animal Protein Complex" or "Global Meat Complex". For example, in the EU, intensive poultry and pig farms (that is, granivores) provide the highest exploitation margin due to production methods that are similar to industrial ones; they eliminate the seasonality of agricultural crops and use cheap grains for feeding pigs and fowls (see Figure 3.2). This intensive livestock model can already be considered as a globalized model that provides cheap meat for both industrialized and emerging countries thanks to the subordination of large amounts of virtual land in peripheral countries that must cultivate grains for animal feed rather than food for their suffering population, and their agroecosystems fall victim of serious environmental impacts (Infante et al., 2018b). This system, which has been growing since the Second World War, forms an authentic industrial complex similar to that of the military industry with extremely high levels of business concentration, as we will see later.

The world's production of meat reached 317 Mt in 2016, Europe and America being the main producers. In that same year, international meat trade reached 30 Mt, almost 10% of production, driven by the Chinese market and other emerging economies such as Chile, Mexico, South Africa, or the United Arab Emirates. Meat consumption, which was 27 kg per capita in 1999, is forecast to increase to 48 kg per capita in 2030 in a context of growing population (GRAIN & IATP, 2018, 3). In the same way, production will be 13% higher in 2026 than in the period used for the calculation (2014–2016), facilitated by two factors: feed grain prices that are low

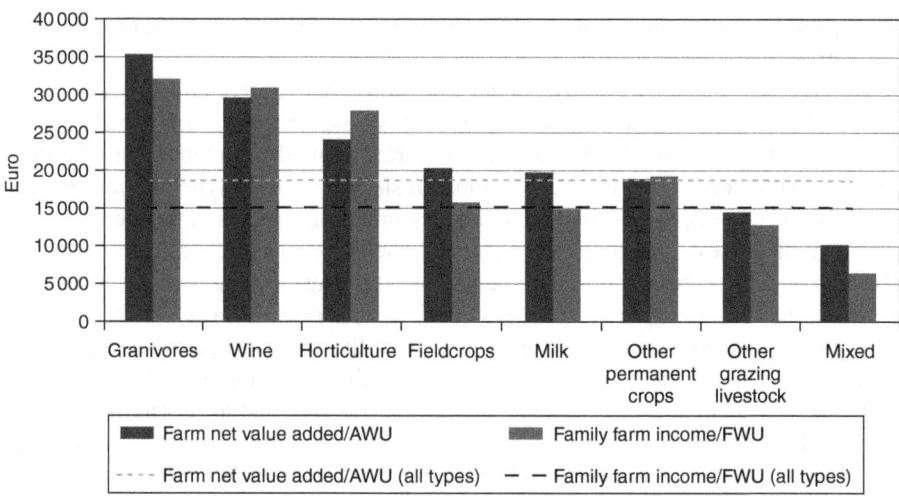

FIGURE 3.2   Farm income by farm type in the EU, 2015. *Source*: DG AGRI, Farm
Accountancy Data Network (accessed November 27, 2018).

already are forecast to fall further or at least remain stable; and nominal meat prices,
which are also low, are maintained and even lower in relation to 2016 (OECD-FAO,
2017, 121). Meat consumption is driven by per-capita income increase in emerging
countries, but this is only part of the truth; it is also driven by the fact that end prices
make meat and dairy consumption very competitive with respect to vegetable production. For example, in Spain, thanks to the low relative prices of raw materials, the
price of poultry and pork is around 3–4 €/kg, the same price as that usually reached
by most vegetable products. Both types of consumer prices have followed opposite
trajectories since the beginning of the 1970s (Figure 3.3) (González de Molina
et al., 2017).

The inefficient nature of livestock production is known. According to GRAIN and
IATP (2018, 3), for every 100 calories of animal feed based on cereals, only 17–30
calories enter the human food chain as meat. This reality is undeniable, and as the
FAO (2006) has repeatedly warned, the use of cereals as animal feed could jeop-
ardize food security, reducing the grains available for human consumption. GRAIN
and IATP (2018) also warn that we must reduce global emissions by 18 billion tons
by 2050 to limit global warming to 1.5°C. If all other sectors follow this trend and
the meat and dairy products industries continue to grow as expected, the livestock
sector could account for up to 80% of the allowable greenhouse gases (GHGs) in
only 32 years' time. Both the emissions generated by intensive livestock farming and
the huge biomass consumption that it represents make meat production increasingly
unsustainable, which is one of the main factors that explains the competition for the
final use of biomass. In this regard, it is directly related to the high prices of food in
poor countries and to hunger. According to a new Greenpeace report (Greenpeace
International, 2018), average per-capita meat consumption should drop to 22 kg in
2030 and to 16 kg in 2050 to prevent dangerous climate change (GRAIN & IATP,

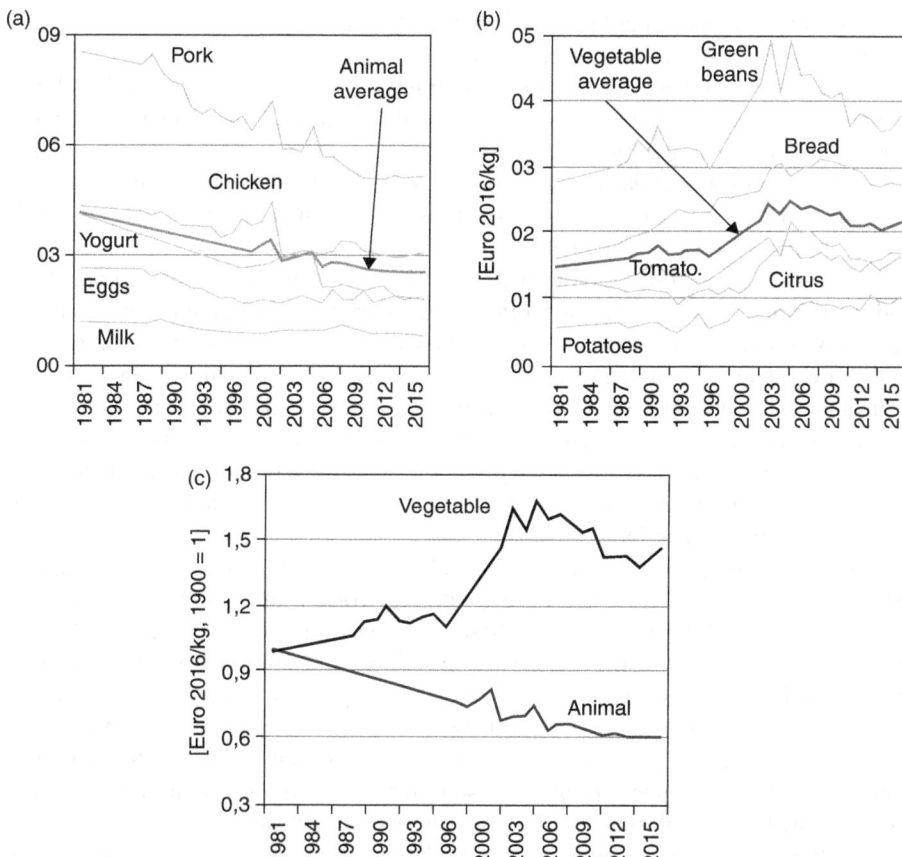

FIGURE 3.3 Comparison between the prices paid by consumers for certain animal and vegetable products in Spain (1960–2015). *Source*: González de Molina et al. (2017).

2018). Clearly, the high consumption levels of meat and dairy products, recorded in industrialized countries and the gradual increase taking place in emerging countries, contribute especially to the unfeasibility of the CFR and is even leading it steadily to collapse.

3.5 "BUSINESS AS USUAL" IS NOT AN OPTION FOR THE FUTURE: LOOKING FOR ALTERNATIVES

There is a growing conviction in the scientific community, governments and think tanks that input-intensive or industrial agriculture is unviable. Its negative impacts on the environment and health, its inefficacy to provide a decent income to farmers, its big reliance on fossil fuels and its vulnerability to climate change have convinced them of the need for a model change. Although enough food is currently produced to

feed the entire world's population, it is not at all clear that the current situation can
be upheld if the biomass produced continues to be unequally distributed, which is
the main cause of hunger and malnutrition. International organizations such as FAO
(2009) recognize that industrial agriculture will be unable to meet increasing food
demand caused by population growth (more than 9 billion people by 2050) and by
the increase in meat and dairy consumption, especially in the emergent countries ...
if things go on as they are, that is. This will raise the demand for food between 70%
(FAO, 2009) and 100% (Tilman et al., 2011).

These percentages illustrate the need for a change of direction in food production
and distribution. However, for supporters of the current regime, the problem can be
solved by technologies that increase the volume of production without degrading natural resources or without aggravating climate change. In this context, some technological models have emerged. Among the most widespread in the current food regime
we can find climate-smart agriculture, precision farming, and sustainable intensification or the "promising way" of bioengineering. None of them intends to change the
institutional configuration of the CFR. Instead, they try to maintain it by increasing
dependence on commercial technologies.

Climate-smart agriculture, termed by the FAO in 2010, is promoted by the Global
Alliance for Climate-Smart Agriculture, with the World Bank, several governments,
lobby groups, and fertilizer corporations participating. Its main goal is that of
increasing productivity using fertilizers, pesticides, and improved seed while controlling and reducing GHG emissions. It is actually a way of pursuing "business
as usual", as it seeks to keep farmers dependent on increasingly expensive external
inputs. The spiral leading to fall in profitability will continue with this type of agriculture and does not appear to produce significant reductions in emissions because
the embodied energy of inputs will not be appreciably affected.

Another agroecosystem management model has been proposed, named *precision
farming*, which promises big changes in crop production. The associated technologies, however, are expensive and therefore only accessible to large farms and capital-
intensive agri-enterprises. A new pack of technologies is included, in particular, those
applying digital technologies to agricultural production; a new generation of agrarian
technologies still at an early stage but developing quickly and closely linked to big
data platforms. Indeed, all the actors in the industrial food chain are developing Big
Data sensors and working with robotics. Agribusinesses using high-flying satellites,
low-flying drones or ground-level tractors to identify crop species, predict yields,
analyze chemical usage, and even determine the patents or licenses associated with
the plant varieties or chemicals. The rise of robotics will therefore not only affect
our way of farming, but also of food processing, retailing, and consumption – that
is, the whole of society. To all this must be added Synthetic Biology (SynBio), also
connected with Big Data, robotics and artificial intelligence, and gene editing technologies to "improve" vegetal food and reduce carbon hoof print of livestock. A third
dimension on which the industrial food chain is focused is financial technologies
(fintech), which includes blockchains and cryptocurrencies, Big Data tools for managing the commercial interrelationships between the food chain actors (Mooney,
2018, 20–24). In this field, IBM, in close relationship with Walmart, developed several
projects that process farm and transforming industries data with the aim of building

a complete history of a specific food. These databases provide massive information about, for example, date and place of planting, treatment dates and characteristics, date of harvest, number of agents the food has passed through, transport duration and conditions, date of entry into the supermarket, storage conditions in the supermarket and, finally, date of sale to the consumer all the way to the consumer's address, information on storage time in the refrigerator and even which part of the food ends up being discarded. Many companies and food chain agents are undeniably interested in this type of information.

In any case, these technologies, rightly called hypertechnologies, are far beyond the control of farmers, especially small farmers, and therefore strengthen the control exercised by large corporations. These technologies enhance the model whereby agriculture becomes a vast market for the input industry and consumers into passive buyers of food products strongly controlled by the agrifood sector and large retail distribution chains. In fact, the development of these technologies has led to powerful mergers, absorptions, and alliances (joint ventures) among large companies that already dominate the food regime.

These new technologies broaden inputs markets that are showing signs of stagnation or uncertain growth. Indeed, traditional input markets are starting to become saturated, especially in industrialized countries. For example, agricultural machinery and technology is huge. With a worldwide turnover of US $137 billion, 2013 was the best year ever for the sector. However, since then, the sales of tractors, balers, milking machines, feeding equipment, and other technical gear have been falling. In 2015 the turnover dropped to US $112 billion. A further decline is expected in the coming years. The European and North American markets are saturated. The number of farms is decreasing, especially in animal production. The area used for farming is shrinking and fewer subsidies are being paid out (Heinrich Böll Foundation, 2017, 16).

Together with climate-smart agriculture and precision farming, a new and more ambiguous term has appeared that has rapidly spread worldwide and is commonly used in the academic literature and reports by international organizations and agriculture think tanks: *sustainable intensification* (FAO, RISE, the UK's Royal Society, etc.). Sustainable intensification refers to a form of production whereby "yields are increased without adverse environmental impact and without the cultivation of more land" (The Royal Society, 2009) and without undermining the capacity to continue producing food in the future (Garnett et al., 2013). The technical strategy consists of making both expressions compatible with each other, that is, the increase in physical productivity would be obtained through environmentally sustainable management practices. To reflect that dual objective, a new mantra has been disseminated: "Growing more with less is the guiding principle, lowering water and land use per unit of output while conserving the productive capacity of soils" (Zhou, 2000, 1).

According to documents of that have had great international repercussion (The Royal Society, 2009; FAO, 2011b; UK Government Office for Science, 2011), sustainable intensification would be promoted through the use of hybrid technical strategies, combining biological–vegetative practices, promoters of the ecosystem's ecological functions, with chemical–mechanical and biological technologies derived from the prevailing agronomic paradigm. Despite its orientation toward an eclectic technological approach, this proposal does not break the structural dependence on industrial inputs or the use of patented technologies and intensive fossil fuel use. Therefore, the mainstream's proposal of "sustainable intensification" has no thermodynamic foundation and thus does not lead to sustainability (González de Molina & Guzmán Casado, 2017). Nor does it advance a critical and more complex view of the power relations that shape the dominant technological paradigm (Loos et al., 2014).

The term "sustainable intensification" is deliberately ambiguous; it intends to cover several supposedly alternative models to the industrial model that correct environmental problems without the need to change the institutional framework that has made it unviable. As stated by Buckwell et al. (2014, 10) in the RISE report: an individual farm wanting to practice sustainable intensification can adopt "one of the farming systems which have been created specifically for their sustainability attributes: Agroecology, biodynamic, organic, integrated and precision farming, and conservation agriculture". In fact, "sustainable intensification is not wedded to any one agricultural approach" (Garnett & Godfray, 2012, 17). As can be seen, the definition tries to include very different or even opposing models of agriculture. Some international institutions consciously promote this ambiguity and even advocate a mix of agroecological and conventional methods: "This hybrid" or "consider all options" approach has been endorsed by major global reports (e.g. FAO, 2011; UK Government Office for Science, 2011; The Royal Society, 2009), which have advocated "the use of ecologically based farming methods without excluding chemical inputs, hybrid seeds, or other management tools" (Garbach et al., 2016, 2–3).

Consistent with this "technical" solution to problems of unsustainability, the preferred way is to use resources more efficiently. Indeed, many think tanks and international organizations that defend the status quo believe that sustainable intensification actually means improving input efficiency, without changing the regime, the industrial farming model, or food markets dynamics: "The prime objective [...] is to improve the efficiency of agriculture" (Buckwell et al., 2014, 28; Lang & Barling, 2012). In other terms, the prime goal of sustainable intensification is to raise *productivity* (as distinct from increasing *volume of production*) while reducing environmental impacts" (Garnett & Godfray, 2012, 14). This ambiguity regarding the farming models tries to maintain the high-input model based on agricultural growth with a new environmentalist language: "Improving agricultural growth is also imperative for reducing poverty, in itself to cause of some forms of environmental degradation and hunger" (Pretty & Bharucha, 2014, 5). More explicitly, some authors suggest sustainable intensification actually justifies new high-input models and the use of technologies, such as biotechnology (Loos et al., 2014) and, more specifically, GMO seeds. But this simply means greening the status quo. "As such, the concept has been endorsed by some interest groups, particularly the farming industry, and criticized by others, particularly those from within the environmental community" (Garnett & Godfray, 2012, 9).

Faced with these proposals that seek to maintain the status quo, parts of the academic world and social movements are proposing an alternative that also contemplates intensification, but that enhances agroecosystems' own available resources, not demand for external inputs, pursuing an "intensification in the use of the natural functionalities that ecosystems offer" (Chevassus-au-Louis & Griffon, 2008). This approach has been called "ecological intensification", unmistakably positioned within organic agriculture, aimed at "maximizing primary production per unit area without compromising the ability of the system to sustain its productive capacity" (FAO, 2009). The term *agroecological intensification* has even been proposed as well. It has been defined as "a management approach that integrates ecological principles and biodiversity management into farming systems with the aim of increasing farm productivity, reducing dependency on external inputs, and sustaining or enhancing ecosystem services" (Garbach et al., 2016, 2). The decisive point is that this approach "focuses on 'natural means' of increasing outputs, for example by incorporating legumes into fields or using agroforestry techniques" (Loos et al., 2014, 2), which eliminates the need for external inputs and, therefore, stops considering the agricultural sector as a market for the input industry, which is precisely what the CFR is trying to avoid at all costs.

Yet, intensification, even if it is ecological, cannot be maintained indefinitely, because it has no thermodynamic foundation. In a specific place and time period, it could be sustainable if the intensification takes place following agroecological criteria. Indeed, Agroecology defends that the only sustainable way to further intensify agricultural production without damaging the natural resources is by using agroecological methods (Gliessman, 1998; De Schutter, 2010; Nicholls et al., 2016); for example, by crop rotations, increasing biodiversity, incorporating legumes into fields, using agroforestry techniques, etc. It could, for example, be the best way to reduce the yield gap that currently exists between conventional and organic production. This gap weakens the possibility of organic farming of becoming a real alternative to conventional production on the horizon of 2050. In other words, only farming managed with agroecological criteria could meet future food challenges in a sustainable way.

Nonetheless, to generally impose an agricultural model based on these criteria, more is needed than the proposal's technological superiority or its social benefits. An agricultural model based on these criteria and management will not be structurally dependent on external inputs. Therefore, it will cease to be a fundamental area to accumulate capital and to transfer income to other sectors. The food regime and the corporations that support it exert strong pressure on governments and public opinion to maintain the same institutional arrangements that guarantee their maintenance. As reported by Olivier de Schutter (2010), blocking factors or lock-ins prevent change: that is, factors directly related to the concentration of power along the food chain, and economic–financial activity generally (IPES Food, 2017, 45).

Two recent reports (Heinrich Böll Foundation, 2017; Mooney, 2018) show the extent to which mergers and alliances of large food companies and agricultural inputs have flourished in recent times. Drawing an analogy with the power accumulated during the Cold War by the "military–industrial complex" – maintained to this day – the globalized food system has been shaped into a "food complex" composed of a small group of large transnational corporations, most of which have intensified their alliances and thus their ability to control sovereign decisions made by States, both nationally and internationally.

Indeed, two notable processes have taken place within the food complex. On the one hand, business concentration has largely increased within the different activity sectors, substantially reducing the number of corporations that control each link in the chain. On the other hand, alliances and mergers also take place between companies representing different links, so a small group of large corporations are expanding their control over more and more links or over the entire food chain. In the same way, alliances have spread to the technology sector-linked Big Data to introduce its applications in the field. Monsanto spent US $930 million to buy Climate Corporation, the agricultural sector's most advanced data analytics company; it has also launched a joint venture with the world's largest enzyme producer, Novozymes (Mooney, 2018, 22). Big Data and intelligent vehicles are making farm production and food retailing attractive for the likes of IBM, Microsoft, and Amazon (Heinrich Böll Foundation, 2017, 11). The concentration and agreement trend between companies has stepped up in recent years and predicts scenarios of near oligopoly in the food industry. Takeovers of Monsanto by Bayer, the mergers between Kraft and Heinz or Dow and DuPont are examples of how the CFR is a playing field dominated by a shrinking number of the biggest players.

According to Mooney (2018, 4–5), four large companies control 67% of the seed market, and the mergers of Bayer–Monsanto, DuPont–Dow and ChemChina–Syngenta are expected to condense the control of the global market in the hands of three large operators. The four largest pesticide manufacturers (DuPont, Dow Chemical, Syngenta, and Bayer) control 70% of the agrochemicals market. Five companies control 18% of the chemical fertilizer market. Three big companies (Deere & Company, CNH Industrial, AGCO) share more than 50% of the global market of farm machinery. Deere alone had a turnover of US$29 billion in 2015: higher than the combined seed and pesticide sales of Monsanto and Bayer (Heinrich Böll Foundation, 2017, 16). A total of 90% of the worldwide grain trade and 70% of all biomass flows passing through it are in the hands of only four companies: Archer Daniels Midland (ADM), Bunge, Cargill, and the Louis Dreyfus Company (Heinrich Böll Foundation, 2017, 26). The world market for processed foods is not yet as concentrated, but the 50 largest food manufacturers account for 50% of global sales. A handful of large companies control the distribution and retail sale of food, for example, four in Germany and five in Spain, and this is the most common pattern in most countries. As the *Blocking the Chain* report comments, "Vertical and horizontal integration continues, but regulators have neither the capacity to monitor it nor the legal tools to control it" (Mooney, 2018, 6).

It can also be distinguished from the "Global Meat Complex" made up of large firms controlling the production, processing, and trade of beef, poultry, and pork worldwide. Cargill, the best known, is the chief supplier of feed grain, the world's second-biggest feed manufacturer and the third-largest meat processor in terms of food sales. Others, like CP Group from Thailand, New Hope Liuhe and Wen's Food Group from China, and BRF from Brazil, are leading feed manufacturers and meat processors in their own right (Heinrich Böll Foundation, 2017, 34).

Due to growing tensions arising from competition for the use of limited agricultural land, expectations that it will increase in the future in a context of growing demand for food and feed, and the consequent increase in prices that can be expected, the production and distribution of biomass has become a business opportunity for banks and other financial firms. Banks such as Goldman Sachs, Morgan Stanley and Citibank as well as pension funds and other investment funds have entered the agrifood market. The market for these new investment products has grown rapidly in recent years. Between 2006 and 2011, the total assets of financial speculators in agricultural commodity markets nearly doubled from US$ 65 billion to US$ 126 billion (Heinrich Böll Foundation, 2017, 38). The effects that the entry of these financial operators are having and will have – to an even greater extent – in the future, i.e., higher prices and greater volatility, have been denounced by the United Nations. The most disadvantaged are those especially in countries for whom food represents the highest percentage of their household budgets (Heinrich Böll Foundation, 2017, 3).

This "food complex" therefore conditions – through lobbying practices, revolving doors, and other less-transparent instruments – governments' decision-making, stealing popular sovereignty and citizens' democratic control. That is why we have been referring to a CFR (see Chapter 2), formed within an institutional framework that favors its interests both nationally and, above all, internationally, that has been built not only on the basis of national legislation, but also on international agreements that guarantee the continuity of this framework and its predominant participation in decision-making. Treaties such as those arising from the successive talks of the World Trade Organization, free trade agreements between different countries and regions of the world, or the most recent TTIP or CETA between the EU, the USA, and Canada are a good illustration of this.

The IPES-Food report (2017, 48) identifies eight interrelated impact fields of this concentration process on the dynamics of food systems: (1) redistributing costs and benefits along the chain, and squeezing farm income; (2) reducing farmer autonomy in a context of "mutually reinforcing consolidation"; (3) narrowing the scope of innovation through defensive and derivative R&D; (4) hollowing out corporate commitments to sustainability; (5) controlling information through a data-driven revolution; (6) escalating environmental and public health risks; (7) allowing labor abuses and fraud to slip through the cracks; and (8) setting the terms of debate and shaping policies and practices.

In the light of the control exercised by these large corporations and the increasingly oligopolistic practices that are being imposed, it seems unlikely that the CFR will encourage structural solutions to the food system crisis based on Agroecology, unless it manages to spread and convert a light or weak version of Agroecology into a hegemonic one, incorporating only some of the technical proposals that leave the core of the current regime unaffected, preserving the input market and its control of distribution. Long-term approaches that reach beyond the specific interests of the companies that control the regime are missing. This means that collapse – and the social suffering that comes with it – is a possibility. Consequently, the only way to solve the current crisis and the risk of collapse is to transform the CFR into a very different one, based on sustainable forms of production, distribution and consumption, democratically organized and controlled.

4 Cognitive Frameworks and Institutional Design for an Agroecological Transition

The transformations that are necessary to build a new sustainable food regime inevitably require a new outlook on the ecosystemic structure of the biophysical reality. As a species, we are fatally and gladly surrounded by that biophysical reality, and we need to build new perceptions in harmony with it. We need cognitive models that guide human action and organization toward generating dissipative structures that replace metabolic entropy with information. To build these structures, we must start by designing and selecting institutions that recover their original evolutionary function: that of reducing social entropy and, at the same time, doing so in a sustainable manner, decreasing physical entropy. As we specified in the first chapter, the most efficient and negentropic institutional models of social interaction are cooperative models.

Consequently, the mission of Political Agroecology is to design institutions that facilitate the changes and adaptations (resiliencies) necessary to move from the corporate food regime (CFR), based on growth, simplification and financialization of its activities, to a new regime based on sustainability. This transition is of a political nature, not only of an economic or technological one, and, therefore, it is not only local but global. This implies changing the systems of material production but also modifying the ways in which public goods related to food are reproduced and socially distributed. These objectives are unviable without changes of a political nature. Agroecological political transformation requires, in turn, a profound cultural change that modifies the dominant cognitive frameworks with which we "construct worlds" (Goodman, 2013) in interaction with the biophysical environment. Agroecology located within neoliberal cognitive and institutional frameworks encounters obstacles and systemic rejection that makes the leap of scale and the agroecological transition impossible.

4.1 THE COGNITIVE FRAMEWORKS OF POLITICAL AGROECOLOGY

The psychologists Hardisty and Weber (2009) have shown experimentally that citizens are willing to pay similar amounts of money for environmental purposes, but they differ in the way they would make those payments: some prefer to pay state taxes and others higher market prices. The payments are directed toward internalizing the negative environmental costs of the fight against climate change in the United States. For some citizens, the same amount of money is unacceptable as a price to pay, but acceptable in the form of a tax rate and vice versa. Those citizens who define themselves as "democrats" are prone to pay it in the form of a tax and reluctant in the form of a price. Conversely, those who identify themselves as "republicans" loathe that same payment in a fiscal form and are favorable if it is reflected in prices. What Hardisty and Weber have shown is that "cognitive frames linked to political or ideological cultures" prefigure and condition the attitudes and behaviors of individuals even above (within certain limits) the economic cost of the decision. The strength of the "political ideology" as a cognitive framework largely determines choices, including economic decisions that are guided exclusively by a type of strict rationality based on the selfish calculation of costs and benefits.

Institutional contexts, ideologically associated, in which the options are presented, condition the choice, the scale of the preferences, and even individual behaviors. This has been known for a long time thanks to the neo-institutionalist school, but it has been verified regarding the cooperative collective management of natural resources thanks to the work of the Nobel Prizewinner Elinor Ostrom. Types of institutions produce structural frames that stimulate or limit individual choices and behaviors. The American political scientist and economist demonstrated two phenomena in particular: that certain types of institutions favor cooperative or community management of natural resources; and that this type of cooperative or community management is optimal for a sustainable management of resources. She shows that there is a structural link between the institutional models and the sustainable management of the territory and natural resources. In *Governing the Commons*, Ostrom (1990) uses case studies and mathematical models based on game theory to show that communal institutions or cooperatives are the ones to have fostered the most sustainable forms of natural resource management over time.

Thus, the political ideology and the type of institution produce "cognitive frameworks" and "behavioral frameworks" that favor some choices and behaviors over others. The fact that a form of election in the political sphere (whether or not to support a particular ecotax program, for example) or an economic one (agriculture, forestry and organic livestock) is based on cognitive frameworks (ideology and institutional frameworks) related to it, strengthens and stabilizes these choices and behaviors beyond circumstantial variations and contingencies. Cognitive frameworks are part of the "black boxes" of ideologies. They shape individuals' perceptions largely unconsciously, to the point of conditioning the system and the scale of preferences, tastes, and individual and social values. Institutions, on the other hand, condition the behavior of individuals through a system of rules, signaling, stimuli, and sanctions that restrict the spectrum of possible alternatives. Political Agroecology, by generating cognitive and institutional frameworks, favors a form of reflexive automation of the perceptions, ideas, and behaviors of the actors and agents aimed at achieving a sustainable metabolic regime.

The expression "reflexive automation" describes the internalization of a critical, metacognitive, efficient decision-making system that avoids or minimizes the temporal costs of reflexive decision-making and the irrationality of automatic or instinctive responses to stimuli or environmental pressures. This stochastic equilibrium, like all ecological balances, between automation and reflexivity, is expressed and embodied through the combination of cognitive and institutional frameworks, of ideology and norms. We will return to this question when we address the issue of neuromimetics. Institutional design thus establishes material and socioenvironmental conditions that encourage putting into practice (behavioral practice) the ideological program that the cognitive framework signals as the preferred or desirable one.

Political Ecology and Agroecology offer a cognitive framework (an ideology) and an institutional framework (a program of norms, actions, and reforms) that favors the integral development of sustainable social production and reproduction. On the other hand, the attempt to consciously or unwillingly insert Agroecology into a liberal ideological framework (ontological individualism[1]) and into an institutional framework proper to neoclassical economics (market) is doomed to failure and to "systemic disgust". We understand that this way of rejecting, expelling or encapsulating a subsystem or a system's foreign object is a form of defense. The "rejection effect" is embodied through the marginalizing of agroecological production and consumption; the expulsion effect implies disappearance or conventionalization; and the "encapsulation effect" involves confining agroecological experiences to a territorial or quantitatively reduced area posing no threat whatsoever to the continuity of the CFR. The greater the ideological and institutional assonance between agroecological practice and the dominant frameworks, the greater the chances of failure, perverse effects and fraud. Consequently, Political Agroecology must undertake a program of reforms and metamorphic changes of the food regime based on two pillars: (a) a scheme of cognitive frameworks, and (b) a scheme of institutional frameworks. For this, it is essential to spell out cognitive principles and institutional criteria that guide the philosophy and practice of Political Agroecology.

4.1.1 COGNITIVE PRINCIPLES OF POLITICAL AGROECOLOGY: IDEOLOGY

We will describe a series of normative principles that need to underlie the cognitive models of Political Agroecology. This cognitive matrix brings about a systemic representation of reality. Like a modern Janus, it has two faces. On the one hand, a political theory of the ecological crisis, and, on the other, an ecological theory of action, relations, and political institutions. The political theory about the ecological crisis is at the same time an ecological theory about politics. The sophisticated nature

---

[1] It is useful to distinguish three definitions of individualism that are not necessarily mutually exclusive. "Ontological individualism" implies an "ontological commitment" and assumes that the community is the sum of individual atoms, as expressed by Hayek: "Society does not exist, only individuals exist" (1944). The second definition is "methodological individualism", which implies only an individual's "epistemological commitment" as a fiction that is instrumental to represent a unit of rational decision-making (be it a human individual, an institution, a collective action group, or an intelligent machine, a recursive algorithm in computational simulations). This definition is typically used in the theory of rational choice and the theory of public elections (Arrow, 1974; Elster, 2007). Finally, "ethical individualism" is a legal fiction that recognizes the human individual as the basic unit of imputation of rights and social capacities (Rawls, 1993).

of the ecological paradigm ensures the intrinsic unity of Political Ecology discourse (Garrido et al., 2007).

This cognitive matrix gathers around three cognitive principles: *biomimetics*, *neuromimetics*, and *etomimetics* (Garrido, 2012). Biomimetics is a classic assumption found throughout ecological thought, especially ecological economy. It consists of learning and replicating the most complex and efficient ecological processes and applying them to social systems. As we saw earlier, agrarian sustainability is associated with the imitation of processes that make ecosystems sustainable (Gliessman, 2007). In other words, natural ecosystems are dissipative structures that reduce both information and energy entropy so from them, we can learn to design or even improve social and technological models. The objective of biomimetics is less that of copying nature, as is commonly understood, and more learning from it.

A study published in the journal *Nature* (Engel & Fleming, 2007) demonstrated that plants' highly efficient transformation of solar energy (photosynthesis) owes to the quantum nature of vegetation's information processing. This empirical confirmation paves the way to the design of hugely more efficient solar panels than the ones we have now. This effectively illustrates how to apply the biomimetic principle: undertaking a twofold initiative of learning and improving solar energy capture technology. In a rough and distorted way, this is what human technology has been doing. An example is how the geometric figure of the circle has been translated into the wheel. Circles are the most abundant geometric shape because their form is the most efficient, presenting the least amount of friction in movement and complete closure. Copying the circle has given rise to the wheel, an essential element in any mechanical device: cars, waterwheels, clocks, steam engines, or bicycles. Producing work or movement from the heat generated by the combustion of fossil minerals is another example of primary and simplistic biomimetics. Thus, biomimetics consists in imitating and improving nature when it comes to rebuilding human production systems, making them compatible with the biosphere (Riechmann, 2006). This cognitive principle would be, therefore, very useful to guide production according to criteria, potentialities, rules, and limits of natural and physical processes. In this sense, both the *episteme* and agroecological methodologies are based on creatively applying the biomimetic principle when designing and managing agroecosystems, for example by: strengthening the density and connective diversity of agroecosystems, or their internal circuits; closing the biogeochemical cycles; taking advantage of the waste efficiently; maximizing the efficiency of the crops in the capture of solar energy, etc.

However, useful information for sustainable designs and efficiency are not only found in the field of production systems. Two other fields of imitation of nature also offer similar possibilities to that of biomimetics: *neuromimetics* (knowledge and information) and *etomimetics* (behavior, institutions, and social relations). Neuromimetics simply consists in using the human brain as a social cognitive model. That is, copying the devices, rules and mechanisms that make the human cognitive system the most efficient and complex machine. The human brain is the most efficient system known. Weighing only 1,300 grams, measuring only 2 m², and consuming just 400 calories, the brain stores more than 100 billion nerve cells (neurons) capable of developing one million synapses (connections between neurons) per second, with a total interneuronal connective density potential of 10^{14} (the highest connective density of an organism in the known universe). By comparison, a supercomputer model designed by IBM simulating the activity of 10,000 neurons consumes 100 kilowatts. Paradoxically, the connection speed between neurons is between 100,000 and a million times slower than the "logical gates" of an artificial computer's silicon. However, the human brain is able to recognize in tenths of a second any of the trillions of sentences (10^4 multiplied by 10) contained in natural language. The answer to this apparent paradox lies in the complexity (neuronal and social connective density) and information (symbolic management, and synaptic plasticity) used in the brain and human society. Some products of the human brain are also a model of extreme efficiency as is the case of natural language or natural numbers: with a finite and small set of symbols (and consuming very little matter and energy) an almost infinite number of mental and social operations (meanings, representations, calculations, predictions, descriptions, etc.) can be generated. The secret of neurocognitive efficiency lies in substituting matter and energy consumption with information, generating order (Morgan Állman, 2003). The brain reveals itself as a model organism from a thermodynamic point of view. Studying the human cognitive system, and generally that of large mammals with a high level of encephalization, allows us to design extremely efficient and sustainable organization and communication systems.

Finally, *etomimetics* allows us to understand our own evolutionary behavior and the behaviors of animals, especially those closest to us regarding evolutionary dynamics (large mammals, great apes, for example). The idea is not that of endorsing any naturalistic fallacy, nor that of converting the ethology of certain animals into ethical codes or of recovering the pseudoscientific falsehoods of social neo-Darwinism (entirely unrelated to Darwin). It is about understanding the behavioral mechanisms that turn our evolutionary history as a species and the behavior of many other species into a model of efficiency regarding the use of resources and relationships with the natural environment. The ethological keys of efficiency reside in the social cooperation between individuals of the same species, in the sober use of natural resources (biomimetic technology) and wastefulness in the use of social relations through games, for example. Both primitive communities and great apes (especially the bonobos) manage natural resources as finite resources and simulate social and emotional resources as infinite (games, parties, duels, sexual rituals, etc.). The objective is to learn what an ecological organization of human relations would be like regarding food production and consumption by looking at the behavior of animals and that of primitive and indigenous communities.

Political Agroecology can use these three cognitive principles (biomimetics, neuromimetics, and etomimetics) as a cognitive guide when analyzing and designing models of social relationship and relationship with the natural environment, capable of maximizing efficiency and sustainability. The potential of this cognitive approach when applied to industrial, rural, or urban development is evident. In fact, many disciplines such as Industrial Ecology, Ecological Economics or Agroecology, are based on these cognitive criteria and, more importantly, their progress in the immediate future will lie in understanding and deepening these criteria.

4.1.2 EIGHT SOLID PRINCIPLES OF COOPERATIVE MANAGEMENT OF NATURAL RESOURCES: INSTITUTIONAL DESIGN

As argued earlier, agroecological practice must be associated with a series of institutional designs aimed at fostering an ecological and equal society. The principles and criteria included in these designs serve as constitutive and operative rules used by institutions that shape and perpetuate socioenvironmental relationship models over time. A concrete example of this type of institutional design is Ostrom's proposal of eight principles to design institutions that cooperatively and sustainably manage natural resources. Based on an empirical analysis of historical experiences, Ostrom showed that cooperative social management of natural resources is more ecologically and socially sustainable than state management or private capitalist management. Agroecology, insofar as it implies socially reappropriating the management of agroecosystems, deeply coincides with this cooperative and community model of resource management. The eight criteria that Ostrom extracted from the case studies are as follows.

(i) The limits of the resources appropriated and managed by the community must be well defined. Cooperative institutions, far from being an area of improvisation or social spontaneity, must strictly delimit the rules, the distribution of costs and benefits and everything related to the legal security of cooperation. We must remember how important it is to have clear and simple rules to ensure the efficient functioning of traditional communities. The communities Ostrom studied were peasant or indigenous communities made cohesive through ancestral customary norms and highly intrinsic motivations (Bénabou & Tirole, 2003; Tirole, 2016). The inertial and simple categorical power of traditional rules must be replaced by intentional and deliberative institutional design.

(ii) Resources must be provided and appropriated in a congruous way. A direct and transparent relationship needs to exist between the contributing of goods or work and the benefits obtained, avoiding free-rider positions or institutionalized parasitism.

(iii) Participation. For the cooperative model to work, rules must be the least heteronomous as possible. For this, it is necessary that participation processes be agile, clear-cut and inexpensive. Participation must be both creative (it generates or modifies rules) and of an auditive nature (it exercises control over existing rules). Informational asymmetry between actors must be reduced to strictly functional limits. The rules must result from the deliberative decisions of community members who must have the competence to modify them. Restrictions and sanctions must be regarded as self-restraints and self-sanctions of the affected actors.

(iv) Monitoring. Decision-making mechanisms related to the evaluation and control of fraud must be reliable, objective, inexpensive, and transparent. Controllers or monitors must be members of the community or be accountable to it.

(v) Gradual sanctions. The sanctions regime should be dissuasive, gradual and internal. Sanctions must operate equally within the community. Sanctions must provide information on the costs of rule violations. It is preferable to manage symbolic or restorative sanctions over strictly punitive or dissuasive ones.

(vi) Mechanisms for conflict resolution. Conflicts must be resolved by the community or collectively. Negotiated agreements should be preferred over resolutions or sanctions. Arbitration and mediation bodies must be community-based and prestigious. Resolving conflicts immediately avoids aggravating them and brings about more satisfactory solutions.

(vii) The state authority must provide a basic, legal recognition of local rights, avoiding paternalism or external protectorates. This implies reinforcing the principle of subsidiarity.

(viii) Family businesses or cooperatives. Actors or companies must follow a model that is directly linked to the territory, collective interests, and future generations.

4.1.3 AGROECOLOGICAL EFFECTS OF COOPERATIVE INSTITUTIONAL DESIGN

These institutional design criteria favor five effects that are significant in any collective, cooperative and sustainable management of agroecosystems:

(i) *The localization effect.* Caring for the equilibria of agroecosystems requires a symbolic (projection of identity) and an economic (expectation of benefits) connection with the territory. Bureaucratic or mercantile centralization damages peasants' cultural and political connection with the territory.

(ii) *The self-containment effect.* Practices derived from applying cooperative rules and criteria generate a peasant or farmer moral economy that stimulates rewarding self-containment, reducing the risks of fraud and free-rider behaviors. This self-contained reduction of fraud reduces the costs of surveillance, control, and sanctions as well as the consequent erosion of community cohesion entailed by the implementation of these coercive instruments.

(iii) *The confidence effect.* This agroecological moral economy fosters trust between peasants and farmers encouraging cooperation beyond the rules. Generally, the incentives and the payment system establish a clear relationship between individual and social responsibility and benefit.

(iv) *The empowerment effect.* The rules lead to a form of emotional praise that strengthens peasants' or farmers' socially powerful identity, not only as a producer of food but also as an environment caregiver and an agent of health and quality of life. The peasant or farmer's social status that deteriorated throughout the modern process of industrialization and accelerated urbanization recovers, with Agroecology a highly modern avant-garde dimension.

(v) *Intergenerational solidarity effect.* Thanks to Robert Axelrod (2006), we know that reinforcing future expectations constitute a powerful incentive for players to bet on cooperative and responsible strategies regarding the deferred consequences of elections. Our described model of institutional design also stimulates intergenerational solidarity by enhancing localization, the community or family nature of farms, and collective participation and management.

Political Ecology provides a cognitive and institutional framework that ideologically strengthens and socially and politically stimulates the development of Agroecology, not as a complementary sector of conventional agriculture but as a global alternative to the production, distribution and consumption of food. Food can be produced ecologically without the ideological and institutional dimension of Political Ecology, but it will be irrelevant ecologically because it will reduce neither consumption, nor the environmental impacts of conventional agriculture to sustainable levels. Without Political Ecology, organic production can only aspire to covering a limited market segment included under so-called "production of differentiated quality". Ideological and institutional empowerment, by intervening in cognitive and behavioral levels, allows a positive feedback relationship between what is believed (ideology) and what is done (behavior), which strengthens the stability of farms and agroecological consumption beyond market (prices) and government action (laws and public money) oscillations.

The approach to complexity is a methodological characteristic of Agroecology, especially with regard to the protection of biodiversity in agroecosystems. Political Agroecology provides an epistemological foundation to the physical complexity of agroecosystems, and, at the same time, it provides institutional instruments to manage social and political complexity. An example of how efficient the interactions between cognitive and institutional frameworks can be is raised in an interesting paper by Altieri and Nicholls (2007) on pest management and agricultural biodiversity. The authors argue that the biological fight cannot be reduced to the individualized "biological control" of pests, but should focus on restoring the natural landscape's biodiversity and adopt a systemic approach. The question is not that of inventing ecological simplicity, but of reconstructing natural complexity. The approach of Altieri and Nicholls is typical of a complex ecosystem approach, in opposition to the mechanistic approach of "biological control". Managing the fight against pests adopting an ecosystemic approach requires an institutional design of cooperative agricultural production management, from the standpoint of both farming culture and institutions (rules, property, etc.).

4.2 AN INSTITUTIONAL DESIGN FOR AGROECOLOGICAL RESILIENCE

We cannot address the agroecological transition without designing a new institutional framework that is resilient to changes in social practices and behaviors; changes that will, in turn, make it possible to gradually transform the current CFR into a new sustainable food regime. Human evolution belongs to a model of evolution called "mosaic" (Barton & Harvey, 2000): it allows recovering old traits but with new adaptive functions, or, to use another term, exaptation (Gould & Vrba, 1982); it allows modifying the adaptive functions of two or more organisms in parallel, otherwise termed coevolution (Margalef, 1993); and it also permits innovative or newly appearing evolutionary traits, that is, emergence (Bunge, 2015). This evolutionary logic also operates within cultural selection bodies such as institutions, in such a way that institutional resilience develops using those evolutionary mechanisms optimally combining exaptation, coevolution, and emergences. A resilient institutional proposal therefore implies: (i) recovering old institutions, providing them with a new adaptive functionality (exaptation) such as, for example, peasant households, communal services, etc.; (ii) modifying institutions that interact with each other, altering in this way their evolutionary functions (coevolution), such as the market and communal goods or these goods and the State; (iii) designing new institutions (emergence) such as cooperative institutions, digital local currencies, agroecological districts, etc.

We know that cooperation has historically been the key to the high resilience of peasant institutions in the face of extreme climate events. Standards of trust were developed through experiences of cooperation that in turn arose with the need to cope with climate risks (Buggle & Durante, 2017). These cooperative institutions survived even after the weather had ceased to be relevant to economic activity. Regions presenting greater interannual variations of precipitation and temperature show higher levels of social trust and of local commercial circuit connections. These areas were also more likely to adopt inclusive political institutions and have been characterized by a higher quality of local governments even today. The findings of Buggle and Durante suggest that exposure to environmental risk had a positive and lasting impact on human cooperation by favoring the emergence of cooperative norms and institutions that mutually strengthen each other. However, they also show that it is only possible to successfully adapt to changes that have occurred (climate and ecological crisis) and induced changes (degrowth and Agroecology) based on a pluralistic and scalable cooperative institutional framework. If the high degree of climatic variability caused networks of cooperative institutions to emerge and consolidate themselves as the most resilient adaptive strategy, the most resilient strategy in the face of current metabolic entropy must also be that of creating cooperative institutional networks. The greater adaptive efficiency of cooperation over other types of coordination is reflected not only in the cultural selection of institutions but also in natural selection. In the coevolution between species, adaptive success is greater in non-conflictive mutualistic coevolution than in competitive or predator models (Northfield & Ives, 2013). In that sense, the isomorphism between social mutualism and ecological mutualism confirms the greater adaptive efficiency of cooperation over competition.

4.2.1 ORIGIN AND NEGENTROPIC FUNCTION OF INSTITUTIONS

As we saw in Chapter 1, the origin and function of institutions, understood as a range of formal (explicit) or informal (implicit) rules (routines, procedures, codes, shared beliefs, practices) are that of regulating the entropy arising in the coordination of social interactions (Schotter, 1981). This perspective has a dual explanatory and normative dimension: entropy describes both institutions' evolutionary origin and their teleological function. All interactions involve an increase in entropy, disorder and loss of work in the transformation process, but in the case of living systems, intentionality, whether it is reflexive or not, adds to this interaction (Dennett, 1996), which implies an increase in complexity.[2] In the case of human social systems, where mechanisms of cultural selection are at work, regulating entropy becomes more sophisticated and

---

[2] We differentiate between reflexive and non-reflexive intentionality to distinguish between a type of action caused by a simple teleological impulse (non-reflexive intentionality) and intentionality that responds to a complex teleological cause filtered or produced by metacognitive mechanisms or self-control (Luhmann, 2010). This difference is also applicable to the type of information produced by institutions.

is expressed in flexible and contingent rules, that is, in institutions. In that sense, the evolution of institutions comprises apparently irrational mechanisms such as religious beliefs, rituals, taboos, or traditions of unknown origin that contain no meaningful or reflexive information. They last because from an evolutionary viewpoint, they play a fundamental role in preserving communities' biocultural memory (Hendrich, 2017).

4.2.2   SCALES AND THE "SOCIAL POINT" OF COOPERATIVE INSTITUTIONS

Returning to much simpler models of coordination is an impossible goal given the social complexity we have reached, but preserving what remains of cooperative mixed information institutional forms is an absolute necessity if we wish to recover the social institutions' negentropic function, much deteriorated by the social division of institutional work. Does this mean having to recover a model of simplified institutions that would be cooperative and gradually scale-up by accumulation of basic units? This unilateral and cumulative strategy is doomed to failure because these simplified institutions do not exist, or if they do, their magnitude is irrelevant, constituting "islands of reduced complexity" that can be interconnected autonomously with each other to form archipelagos of minimum complexity without any relation to the prevailing "deep institutional structure". Critical thresholds (exogenous criticality) apply to the growth of institutional forms when they are antagonistic to the dominant institutional framework. These thresholds can only be overcome through cooperative multilevel collective action or, to put it in another way, through political action when it is oriented, not toward producing local, endogenous and self-applied changes, but global systemic changes exogenous to the actors themselves.

It has been possible, through experimental trials, to identify these critical thresholds, and overcoming them requires global social changes. In an empirical study based on a methodology from experimental economics, researchers from the Universities of London and Pennsylvania identified a "social point" as a barrier to change (Centola et al., 2018). They experimentally tested a theoretical model for 10 years identifying the "social point" or point of inflection at 25% of the necessary critical mass, above which a hitherto minority new institution, rule, belief, or convention becomes a majority one. The threshold represents a sort of quantitative barrier to social hegemony. From that threshold onwards, an idea, opinion or belief expands and becomes an element of "common sense". Surprisingly, this algebra of social change is only feasible if it operates in a neutral space where there are no powerful incentives to reject change, but beyond the powerful empirical evidence of this model, what is undeniable is the existence of a scalar "social point" that makes systemic change impossible if it is not overcome.

It is unlikely that agroecological organizations at basic scales (at a farm scale or a local, cooperative, or district scale), expand and reach this "social point", i.e., that they escalate or massify or change of scale if they do not resort to an institutional, cooperative, democratic, and mixed information design through multilevel collective action. This implies intervening at the most immediate and close level (farm) but also on a more complex plane (the State). Without this leap of political scale, without neighboring this "social point" it will be virtually impossible to change the food regime and the metabolic regime. Circumscribing collective action to localities or to "archipelagos of hubs of minimum complexity" implies allowing agroecological experiences to be more vulnerable to social entropy emanating from a highly entropic dominant environment. Either one intervenes within that political and state environment, or that very environment, beset with a high level of complexity, eventually inflicts chaos, and the communicational noise expands until it neutralizes the micro-institutional framework of the agroecological experiences. The "hidden institutionality" expands the social division of institutional work and generates a range of non-explicit normative prescriptions that govern the management and design of institutions in a non-cooperative way. Multilevel agroecological collective action must be oriented toward a dual objective: on the one hand, it must intervene in strengthening or building local cooperative institutions of mixed but highly reflexive information; and on the other, it must act on a more complex scale (political/state) as a social movement that intervenes in conflicts with the objective of modifying the hostile institutional environment. From ways of managing farms to social movements, from local markets to city councils or regional governments, multi-level collective action enables extending the diversity of institutional frameworks informed by cooperative coordination designs and practices.

4.3 DIVERSITY OF BASIC-SCALE AGROECOLOGICAL INSTITUTIONS (THE FIRM)

According to conventional theory on firms, the correlation between a firm's institutional design and the mode of production, between the rules and institutional routines and the impact on social inequality or social metabolism is ignored. Firms' asymmetric internal structures have been systematically disregarded, just as the radically asymmetric institutional environment frameworks have been rendered invisible by general equilibrium theories. Power is not an object of study in neoclassical economics (Anisi, 1992). This correlation, however, has not only been overlooked by neoclassical economics, but also largely by Ecological Economics. If, as we have explained, there is a constant interrelationship between institutions and the environment, it is not possible to decouple institutional form from its impacts on the economic, social, or metabolic environment. In this way, institutional design becomes exceedingly relevant in Agroecology and especially at the most basic scales, such as farms.

For this reason, it is worth developing an institutional typology of models that are more adapted to the demands of farms but without directly addressing the property regime: this way, it is possible to sidestep the diversity of national regulatory contexts. However, we must remember that an optimal property regime would be one that is closest to traditional communal forms of ownership and management of goods. This implies an ecological re-reading of the very concept of property in the sense of ownership of use, usufruct or trusteeship (Garrido, 1998).

4.3.1 THE FAMILY INSTITUTION AS A PREFERENTIAL AGROECOLOGICAL FIRM

The democratic family institution, in which men and women participate equally, constitutes the optimal model of agroecological institutional design at a basic scale (the farm). A household, as an institution, with an internal cooperative configuration, where time, costs, and work are equally shared as well as decision-making rights, is key to preventing family institutions from incurring similar asymmetries to those of conventional commercial firms, which generate a highly entropic model. In that sense, the role of the ecofeminist movement and women is fundamental, given that family and reproductive work is a battlefield of ecofeminist struggles and practices. The influence of a household economy is huge and the major part it plays is confirmed by the scientific literature. Four aspects highlight the role of households as an economic institution: (i) the institution generates very strong trust dynamics within its interactions; (ii) it provides stability and autonomy in the face of turbulence in a globalized financial institutional environment; (iii) it facilitates interconnections between culture and the economic environment; and (iv) it is the most widespread form of institution in the world economy (Alesina et al., 2015; Amore & Epure, 2018).

From an agroecological perspective, the preferential role of households is due to a range of factors: because a family institution stimulates intergenerational solidarity through intentional non-reflexive types of information (Axelrod, 2006) as well as reiterating and continuous interactions over time (Garrido, 1996); it increases the intertemporal discount rate, thanks to shared future expectations (Gintis, 2006); it fosters biocultural memory connections and continuity (Barrera & Toledo, 2008); a household stimulates community connections and socioenvironmental capital by establishing and protecting spheres of non-commodified and mutually supportive activities; it favors integral connections between reflexive and non-reflexive interactions, production and reproduction, individual and collective identity; and it brings about and stimulates performative behaviors whereby action and results are intertwined in a "common way of life".

4.3.2 A COOPERATIVE INSTITUTION AS A PREFERENTIAL MODEL OF AGROECOLOGICAL FIRM

To this effect, there are two types of cooperative firms: *endogenous* firms, in which the impulse for cohesion and cooperative coordination is internal or historically and community-based, for example, the customary community institutions studied by Ostrom in *Governing the Commons* (Ostrom, 1990); and *exogenous* firms, in which the impulse for cohesion and cooperative coordination comes from the outside and is contemporary. These institutions are formal, regulatory, as well as regulated and individuals take part in them voluntarily. These types of cooperatives can be classified under Parson's community/association binomial, as egalitarian voluntary associations (Parson, 1976).

In the endogenous model, non-reflexive information carries a much heavier weight than in the exogenous model, where reflexive normativity is far greater but its performative power is much weaker. In any case, due to the processes of cross-cultural hybridization, both endogenous and exogenous cooperative forms are condemned to mixed formulas of reflexive and non-reflexive information. Ostrom highlights this when she points out that community management based on customary practices and norms requires exogenous formal institutions with broader and more complex legal frameworks that protect the autonomy of these forms of community management.

4.3.3 LOCAL MARKETS

The history of mercantile institutions is rife with self-justifying myths, far removed from the original history and nature of what we call "markets": in reality they constitute an institution among others that regulates (coordinates) the social interactions generated by the exchange of material goods. Service, labor, or financial markets are much more modern creations and result from the regulatory and coercive capacity of the State (Polanyi, 2001). Initially, markets worked and established themselves as very primitive cooperation mechanisms based on one of the simplest forms of cooperation and altruism: direct reciprocity (Gintis, 2006; Nowak, 2006). In their early stages of development, markets were "economies of exchange", distinct from "economies of terror" (State) and the "economies of love" (donations) (Boulding, 1994). With the increase of social complexity, the social division of labor and the mediation of money as one of the State's political instruments, the market was transformed into an exchange institution of increasingly abstract title deeds dependent on the state "terror economy". In turn, the "love economy" (Mauss, [1925] 2011) was relegated to local and family economies and has continued to have some weight in indigenous and peasant economies (Toledo, 2017). From the perspective of the agroecological transition, it seems appropriate to recover the evolutionary and cooperative functionality that underlies these three types of economies: cooperation through direct reciprocity (market), indirect reciprocity (donation) and multilevel selection (State). This process of exaptative recovery requires deep and complex scalar democratic governance centered on the federalist principle of subsidiarity.

The market institution thus dates back to long before capitalism and the industrial metabolic regime. Nothing antagonizes financial agribusiness global markets more than local or autonomous peasant markets. Local markets of proximity and peasant fairs as a non-capitalist form of exchange are, and must be, a very useful institutional playing field in the agroecological transition. The objective is to conserve, recover or build local markets of proximity to contribute to: building endogenous networks of development and valorization of peasant production; short marketing channels that reduce the carbon footprint and the consumption of materials and energy; institutional diversification of types of exchange (exchanges of time or services, commercial exchanges, donation activities); democratizing the agri-food regime, given that local markets allow direct decision-making by farmers and consumers, converting prices into a transparent information system; diversifying production thanks to the non-specific demand of the local population's food needs and to the disconnection of hyperspecific demand from global markets.

Product diversification fostered by local markets has an impact on the reversal of genetic impoverishment and on the conservation and recovery of native varieties, consequently increasing genetic biodiversity. Local markets encourage consuming seasonal foods with pedagogical consequences on food habits and demand by synchronizing production with the population's metabolic biorhythms. Therefore, they improve diets and channel demand toward more sustainable consumption. They constitute a practice where the pact of cooperation between citizens, producers, and consumers can be sealed. This pact needs to sustain and promote the social movement of politicization of food consumption, a key component of the food regime's political change. In short, local markets reinforce food security and autonomy, while decentralizing production and consumption.

Local or peasant markets, in their role of exchange institutions, thus represent a powerful institutional instrument in agroecological transition strategies. They represent one of the agroecological movement's most commonly used instruments. Yet, if we wish to prevent them from becoming socially insignificant, or, what would be even worse, from complementing the dominant food regime as they are reduced to niche markets, local markets must be the object of supralocal political interventions by democratic public powers. Democratic public intuitions must protect the autonomy and democratic and cooperative self-organization of these markets by means of public policies. This means using fiscal and regulatory instruments as well as providing public services (public spaces, collection centers, information and transaction platforms, ecological diets in schools and hospitals, etc.; see Chapter 7).

Without the intervention of the democratic State, the agroecological movement would be an isolated area of resistance, converted into a "beautiful green complement" in the midst of an agro-industrial slum. If we examine the obstacles to the escalation of agroecological production and consumption, we find what we call a "systemic effect of rejection". This rejection cannot be removed without external political intervention. The factors at work, which are comparative disadvantages for Agroecology, originate in the systemic environment as well as in the structural economic and political conditions and their effects on agroecosystems, for example: costs related to time, to transactions (certification); abstinence from the use of toxic agrochemicals and their effects during the reconversion process; the impossibility of closing cycles due to mercantile and fragmented territorial planning; owing to the necessary agroecological cultural capital (training and education); to the very complexity of design of ecological agroecosystems, etc. In a different political and systemic environment, these same factors that are barriers in this environment would turn into opportunities.

However, the local or peasant market institutional model does not solve the problems that arise at the different scales in which the regime operates. The principle of institutional diversity, which we have proposed based on the well-known thesis of Ostrom (2013), becomes especially valid when it comes to facing resilience strategies amid the turbulence affecting food autonomy and sovereignty in the current CFR crisis. This institutional diversity involves preserving and accumulating spaces that are disconnected from the global dynamic and that can serve as replicable experiences (socio-technical innovation) in the not-too-distant future before the progressive "nutritional vacuums" that the climate crisis, the ecological crisis, and the food crisis are causing. It is neither our wish nor our goal to design a closed, complete and perfect theoretical model of what agroecological marketing networks would be, but it is useful to describe concrete institutional models, with ample accumulated social practice, that can serve as an orientation, as a signaling system (Schelling, 1989) in the agroecological transition.

4.3.4 LONG CHAINS OF AGROECOLOGICAL TRADE

The long chains of commercialization of food production are highly inefficient, both socially and ecologically. The fact that agroecologically produced food circulates does not prevent it from incorporating high levels of consumption of materials and energy. However, an agroecological transition strategy cannot ignore, beyond its dominance, its hegemony[3] in this globalized regime. No rational and reasonable strategy of agroecological transition could consider that local institutions represent instant and full-blown models of substitution of the food regime's institutions and global markets. The fact of characterizing agroecological institutions as resilient (exaptative) institutions rather than alternative or substitute ones means we have to propose reforms to global markets and long circuits of commercialization.

Three lines of reform can be advanced in accordance with the Ecological Economics, which would help to reduce the weight of the current model. They should serve as guidelines for a future sustainable regulation of international agrifood trade. The first line is that of promoting the internalizing, through fiscal instruments, the environmental, social, and health costs of agriculture and conventional livestock in order to reduce the comparative advantages they today enjoy. The second is to combine short, medium, and long circulation channels, creating limited-duration hybrid circuits during the transition process toward sustainable channels and, therefore, close channels. They could be classified fiscally according to energy indicators (EROIs) and their carbon footprint applying progressive tax rates. The third line is to restrict intra-industry trade and regulate interindustry trade, recovering the classic distinction made in political macroeconomics (Krugman & Obstfeld, 2010) between interindustry trade, focused on the exchange of non-homogeneous goods between bioregions, that is, on products based on the singularity of the productive factors (specific endowment of resources such as climate, varieties, soil, or agricultural habits), and the intra-industry trade through which similar productions are exchanged (homogeneous goods). Long-distance trade would only be justified, and with regulatory restrictions, in the case of interindustry agricultural and livestock trade.

Such measures would only be fully effective in a context of international democratic policies and institutions disposing of true regulatory competence to regulate international trade. Political Agroecology is therefore interested in maintaining and strengthening policies and multilateral institutions as well as their progressive democratizing. Conversely, unilateralism in international politics, developed since the 1990s by the world's agroindustrial powers, especially the USA, and the tendency imposed by the CFR to privatize public international law on trade through agreements and trade treaties between States, seriously undermines the prospect of ecological and fair regulation of world trade.

Designing global democratic regulatory institutions entails enormous theoretical and practical difficulties for international actors; but the ecological and social need to move toward a global state of law is evident; we cannot take refuge in local comfort zones and ignore the situation, advocating a road map based on a simple cumulative spread of local agroecological experiences and institutions. No, that strategy is tedious and idealistic. It is doomed to failure for two obvious reasons. The first is well-known in the social sciences: social changes operate by qualitative leaps (emergence) of scale. There is no time to wait for a presumably very long process of

---

[3] The imposition of the current food regime does not come exclusively from States' and multinational corporations' direct coercion of norms, institutions, and rules (domination), but also from the cultural implementation of preferences and demands that are instrumental for the CFR and are currently hegemonic. The "cultural wars" or the battles between antagonistic cognitive frameworks are essential to agroecological political mobilization because that is where the battle for hegemony takes place.

change based on the mushrooming of experiences; the agri-food and ecological crisis is taking place much faster than the social changes necessary to produce a sustainable political system. In this sense, the propagation of intermediate supranational political structures such as, for example, the European Union or Mercosur and Unasur (Bermudez Gómez, 2011; Rosset & Altieri, 2017) are greatly significant. These intermediate supranational political structures with all their current and numerous imperfections constitute an institutional laboratory of what a cosmopolitan democracy and State can be, while operating as de-facto normative and political actors able to somewhat limit the power of transnational corporations. The international agroecological movement should also support and promote reforms to these supranational intermediate structures.

4.3.5 AGROECOLOGICAL DISTRICTS

Agroecological districts are institutional networks of supralocal and interlocal cooperation; they can be assimilated into regions or bioregions in order to generate forms of cooperation and democratic coordination based on collective decision-making, with the goal of optimizing natural resources from an agroecological, technological, and institutional perspective. Agroecological districts represent an initial attempt to integrate agroecological production and consumption on a supralocal scale, thus incorporating a greater degree of complexity. Finally, agroecological districts include an institutional area of co-production between the State and social agents to generate public goods on a scale of medium complexity. Nevertheless, an optimal determination of agroecological districts does not necessarily imply mechanically demarcating bioregion criteria or biogeographic zoning (Vilhena & Antonelli, 2005). Agroecological districts are an institutional land-zoning innovation aiming at favoring the integration and efficiency of agroecological production and consumption, rather than descriptive maps of ecosystemic specificities. Agroecological districts intend promoting institutional contexts that are friendly and conducive to agroecological production, trade, and consumption on a supralocal scale. They are as unexplored as they are necessary for Political Agroecology.

Agroecological districts, in addition to bringing about economies of scale and scope, such as the mutualization of services and other advantages, can help an agroecological political culture to emerge. The districts' organizational complexity stimulates and trains forms (institutions) and skills toward abstract cooperation that goes far beyond networks of immediate, face-to-face cooperation, based on direct reciprocity or donation proper to institutional spheres such as (family or cooperative) farms or local markets or peasants. The crisis of the global food regime cannot be tackled by adopting an exclusive strategy of simplification toward a local dimension and direct reciprocity institutions. In the following chapters we will discuss how these districts can be implemented by the creation of local agroecological-oriented food regimes.

4.3.6 VIRTUAL LOCAL CURRENCIES

Community currencies have a range of virtues that facilitate recovering the informative dimension of money while deteriorating its speculative and entropic dimensions.

Community currency is both a social movement (that is cooperative and autonomous) and a community institution of agreed social value conversion as well as a method of putting a price on social value. Its adaptive formal plasticity allows it to function or not, depending on the political conjuncture, as a complementary legal currency and to be a pilot model that can be extrapolated to state currency production. It guarantees collective (democratic) control of the monetary system. It allows operating with null interest rates. It facilitates initial credit exclusively filtered by social and environmental interests. In its first stage, it is born out the free will of the participants adopting the form of a credit cooperative; as a result, it has very low original transaction costs. Participation in the initial capital fund does not have to be monetary or patrimonial. It has a simple and transparent participation and control mechanism. It limits accumulation by means of currency oxidation (temporary limitation of validity) and by the regulated limitation of capital rates. It presents no inflation and deflation risks because it is indexed to the community's production of goods and services. It is a proximity and neighborhood currency that promotes endogenous relations. It allows joint coproduction and co-management between social movements and municipal and regional democratic public institutions. Finally, it involves a non-traumatic, partial, and gradual disconnection from the global monetary system because it tolerates partial connections, such as micro-credits within the mutual credit system.

4.4 DEMOCRATIC GOVERNANCE AND DIFFUSE STATE FOR THE AGROECOLOGICAL TRANSITION

The governance concept arises as an attempt to harmonize the relations between the State's public powers and social actors in such a way as to change the traditional relationship of coercive hierarchy in favor of horizontal collaboration. This occurs within the framework of the Welfare State and the growing democratization of both public administration (transparency, participation) and private activity (gender equality, democratic households, co-management in companies, etc.), where the growth of public services such as education and health are forging a democratic intermediate space between civil society and political society. The proletarianization and massification of old and minority liberal professions (health, education, law) has brought about a major sociological change that has created links between the social and democratic state and the social actors that have changed the political culture (Offe, 1990).

The State's growing socialization and the progressive democratization of the so-called civil society requires building non-dichotomous political categories (political society/civil society) that bring democracy and the production of public goods to share common concepts, understood as a shared function between the State and social actors.[4] Neoliberal counter-reform rose precisely against this incipient idea of democratic governance that threatened the status quo of capital. It is precisely that neoliberal reaction that has seized the idea of governance and used it as

---

[4] The distinction between political society and civil society has been criticized, and rightly so, as it conceals class inequality as well as the connections between classes and the State. We prefer to talk about social actors and thus include social movements as public agents. In Chapter 7, we will see how this consideration is applied to the requirements of co-produced public policies.

a battering ram to weaken the State and democratic social actors such as unions and social movements. Boltansky and Chiapello (2002) described it in "The New Spirit of Capitalism" as a neoliberal strategy of resignification and capture of antagonistic arguments and semantics to use them in a diametrically opposed sense. While democratic governance meant cooperation between public actors and social actors in the coproduction of public goods, neoliberal appropriation has led private actors to penetrate the public sphere to produce goods and private income.

It is thus urgent to grab the concept of democratic governance as it establishes a new relationship between the State and social actors based on a new horizontal and cooperative model between democratic actors. To address the social and environmental crisis in which we are immersed, we cannot validate public initiatives that are alien to behaviors, beliefs, and social institutions, as in the case of the bureaucracy of real socialism or authoritarian states; nor is neoliberalism's chaotic privatizing deregulation valid either. A new institutional framework of cooperation between State and society is needed, whose wheels are greased by democratic flows. Bureaucratic statism increases political entropy to untenable limits reducing social entropy, but it ultimately increases metabolic entropy. Conversely, liberal privatization increases social entropy, decreasing political entropy but also finally increasing metabolic entropy. Real socialism and neoliberalism ultimately make physical or metabolic entropy grow, making the other two types of entropy rebound, decisively.

The evolutionary and ecological theory of institutions developed throughout this text helps us to put on hold the ontological abyss that we assume to exist between society and State. Democratic governance allows social and public actors to manage metabolic entropy comprehensively and in a coordinated fashion. Democratic governance avoids the risks derived from an "institutional monoculture" that Olson (1971) warns us against. In an era of collective intelligence, it makes no sense for a group of experts or a group of bureaucrats to concentrate all decisions, impoverishing the enormous wealth and creativity that is dispersed over so many social nodes and sensors, capable of processing volumes of information at an unprecedented speed.

4.4.1 AGROECOLOGY AS COLLECTIVE MULTILEVEL ACTION

The agroecological movement is a social movement that self-organizes around a rationale of collective action. It is not immune to the multiple costs and conflicts that this entails. When we speak of collective action we refer to forms of individual action coordinated in a voluntary and cooperative way whose purpose is shared by all those taking part in the action. Therefore, we are not referring to coordinated actions that individuals have been coerced into or that are imposed by an external regulator. In this sense, the agroecological movement is a social movement sustained by ancestral peasant and indigenous practices and knowledge, by scientific research or by movements of ecological criticism of economic growth (Toledo, 2012b), among others.

Like all social movements, however, the agroecological movement is experiencing the problems common to all collective actions. For example, the costs and perverse effects of the intentional coordination of individual cooperative actions described by Mancur Olson in the Logic of Collective Action (1971) or the "transaction costs" described by Coase (1994). These transaction and coordination costs are "noise" (entropy) that hinder and reduce the efficiency of collective action. According to the theoretical framework described in Chapter 1, we could consider them to be the consequences of the social entropy generated by individual interactions in complex social contexts with collective objectives and patterns. As explained earlier, institutions have precisely the function of managing and reducing social entropy levels, in such a way that public goods are produced in the most efficient way possible.

The use of a biomimetic criterion, derived from the theory of evolutionary multilevel selection, understood as the synchronic action of natural selection at two levels, at least, of the evolutionary hierarchy (Okasha, 2006), can help us to understand the original nature of the agroecological movement and define an effective instrument for the agroecological transition: multilevel collective action. It is this latter type of collective action that synchronically intervenes, modifies, at two levels, at least, of the social complexity scale,[5] for example in farms and in the locality or nearby peasant community.

However, multilevel action can be approached in synchronic or diachronic terms. As in biological selection, the relevant point is not interaction on different scales, but whether the collective action is self-conceived and self-programmed to be synchronous, on different scales at the same time, or diachronic, at cumulative scales over time: first the farm, then the local community and so on following a sequential, mechanistic, and cumulative chart. Whether to make its change efficient or because of the very nature of the movement, multilevel collective action must be self-programmed to guide its objectives to as many scales as possible and in a synchronic manner. This implies that agroecological collective action must also be oriented toward designing and implementing public policies, especially those carried out by the State. The dichotomy between social action and institutional political action, between management and social mobilization is an intellectual and political trap in which the agroecological movement would avoid. Multilevel collective action does not deteriorate any of the levels in which it operates, it multiplies the efficiency of the efforts. A reverse approach condemns one part of the agroecological movement to isolation in local areas or farms and the other part to powerless bureaucratic institutionalization.

The evolutionary and ecological theory of institutions we have considered leads us to approach collective action at different scales and within the various institutional frameworks intervening in the production of public goods. Given the current state of social complexity, collective action cannot exclusively focus on a single scale. It can be conceptualized in a restricted manner as the result of the coordinated actions of multiple individuals sharing one or more common purposes. Any action can thus be understood as an impulse to transmit information (negentropy) or to transmit noise (entropy). From this perspective, the mission of multilevel collective action is to introduce information at different scales through various circuits: autonomous circuits for the production of local public goods (farms, households, local markets,

---

[5] In this case we prefer to speak of "scales of social complexity" rather than of "social hierarchy", because the first expression refers directly to levels of complexity, connective density between knots or elements, rather than to dependency or dominance.

virtual social currencies, production cooperatives, services, districts); exogenous circuits of normative social innovation: conflicts and mobilizations, alliances with other actors and social movements (consumers, health, environmentalism, labor unions, solidarity economy, feminism, anti-racial); cooperation with the scientific community; and also state co-production circuits of public policies at the different levels of the State's public administration. These three information circuits correspond to different scales of multilevel collective action that range from the most basic scale of the farm, to the intermediate community scales or the political scales of the State. The agroecological movement has the power to generate a synergy effect and positive feedback between these three circuits of action (information) to generate scale changes in the agroecological transition.

4.4.2 Normative Action and Popular Sovereignty as a Procedure

Efficient cooperation and coordination between social actors, between social movements and political actors (public institutions) lies at the heart of the governance democracy we have been describing, which represents a desirable normative horizon aiming at fostering an agroecological transition that would be the least traumatic as possible. So far, we have gone into the detail of an optimal institutional design of agroecological micropolicies and we have deliberately overlooked the macropolitical State level. This chosen sequence of analysis seeks to reproduce a similar road map to that followed by the agroecological movement until now. The time has come, however, to address the question of the State and political actors, a topic that an important part of the agroecological and peasant movement is reluctant to address, given its original communitarian and self-managed tradition. This political rejection of the State resides not only in cultural or ideological traditions, but also in the war that the State has waged and is waging against the indigenous people, peasants or against the agroecological movement itself. In many countries, Agroecology has survived *despite* public powers. Unsurprisingly, many agroecological, indigenous people or peasant activists regard the State more as an enemy or a danger than a potential ally.

If collective action is one of the core functions of social actors, the characteristic function of political actors is normative or regulatory action. In States, the regulators are always external to the behaviors being regulated. Collective action is self-regulated, normative action is hetero-regulated. The State regulates through legal norms and the issuing of currency (Luhman, 2009; Bicchierri, 2016). The normative capacity is legitimized by a core concept coming from Roman and medieval political theology (Kantorowicz, 1985; Aganbem, 2006), that is, the concept of sovereignty: the supreme power over which there is no other power. The original social contract of modern States focused on property and the subject as a property owner. Establishing a legal security regime to protect property was the first consensus that justified the voluntary self-limitation of individuals' power of autonomy. This self-defeating social contract was justified, in the absence of a theocratic foundation of power, by the need to avoid the "state of nature" (Hobbes, 1984) and its resulting violence and widespread insecurity. The democratic development of this contractual legitimization of political power sought in the construction of a collective subject (the people) the new body of the dethroned sovereign. However, the characteristic that defined the concept of sovereignty was that of absolute power, something or someone nobody is above. With popular sovereignty, the people were the new sovereign, but the absolutist origins of this concept created contradictions in the democratic discourse. The fact that the people were embodied in a "general will" in Rousseau, or in a nation (Sièyes), foreshadowed a series of totalitarian and authoritarian deviations. Collectivist organicism or the despotism of the majority are historical examples of these deviations. An imposing conceptual, institutional construction has been erected against the authoritarian perversions of popular sovereignty. Constitutionalism and guaranteeism have attempted to build a network of powers and counterpowers, of constitutional limits to guarantees and fundamental rights aimed at shaping an invincible barrier to totalitarianism and despotism.

To face the challenges imposed by the agroecological transition, we need a new system of values and principles entailing the expansion of the moral community in such a way that it includes the totality of the biotic community and future generations (Singer, 2002). The need to include them within the sovereign subject is essential from a sustainability perspective because they have neither action nor word; owing to their very condition, their seat is empty. Only by extending the limits of the moral community, beyond our generation and our species, can we make ethical and political commitments capable of rejecting growth to avoid succumbing to the most undesirable scenarios of the ecological crisis.

To address these challenges, we must find a way to limit sovereignty that goes beyond liberal and republican formulas, without forgoing the guarantees that these have provided. This can be found in Habermas' proposal of "popular sovereignty as a procedure". The original idea comes from the German philosopher Julius Fröbel who proposed to replace the substantialist content of popular sovereignty by a series of procedures that would guarantee the deliberative and rational formation of opinion and public decisions. The pragmatic conditions of rational deliberation must be constitutionally safeguarded as a condition for the possible exercise of sovereignty. These pragmatic conditions of communication coincide with democracy's ethical and political intuitions (autonomy, liberties, equality in access to public space, search for truth, common good and consensus, etc.).

The proposal consists of reformulating popular sovereignty as a constituent and constitutional procedure, by which individuals' autonomy is guaranteed to be exercised continuously and permanently, including members of future generations, in equal rights and conditions. This diachronic obligation to preserve the rights of all generations entails a form of subrogated obligation to also preserve the rights of the entire bioethical community, without which the protection of future generations would be ecologically impossible.

4.4.3 COOPERATIVE DEMOCRACY

If cooperation and democracy are evolutionarily efficient forms of organization of the human species, why has democracy been such a scarce political regime until the twentieth century? Where were these evolutionary trends in the broad historical period between the hydraulic empires and the French Revolution? As we have already seen, that period is not truly an extensive one compared to the thousands of years of existence of our species. However, let's examine how modern political philosophy has understood this anti-cooperative parenthesis. According to Rousseau, the appearance of forms of private property marked the end of the "state of nature" that, for the Geneva philosopher, contrary to Hobbes, was not the kingdom "of all against all", but of "all with all". Private property was the original sin that drove us out of primitive community paradise. Rousseau understood the social contract as an institutional way of recovering natural freedom through political freedom. Individuals do not endorse the social contract to protect themselves against the threats of the "state of nature", as is the case for Hobbes, but to recover the freedom that was in the original "state of nature". The "original sin" of private property has destroyed the community with the formation of the State and social classes, with the fragmentation of the unity between politics and community.

The destruction of that unity between politics and community is the result of a jump in demographic, political, and technological complexity, marked by the passage from the Paleolithic to the Neolithic and the appearance of major agricultural surpluses that had to be managed. Money emerged as a universal symbolic form, from the technical need to manage the complexity of the exchanges, as a political instrument of domination and generation of inequality. This process resulted from dissolving the closed and circular metabolic regime of primitive communities. That very rupture is also the remote cause of the ecological crisis. With the beginnings of agriculture, social cooperation gradually deteriorated. Inequality and political division began to grow. Money and unsustainability joined hands in generating inequalities and deep social asymmetries. Different sorts of social metabolism condition different types of organization (González de Molina & Toledo, 2011). A circular metabolism maintained cooperative and egalitarian forms of organization for thousands of years.

Democracy was not born with modern parliamentary systems; countless cultures developed egalitarian, cooperative and participatory systems of social organization long before the eighteenth-century revolutions (Guaman, 2015). Although the modern recovery of democracy has unfolded at the same time as capitalism, this common path has become increasingly adversarial because of the disproportionate increase in inequalities and the ecological crisis. Economic growth allowed a relaxation of the coercion systems and empowered social actors struggling for democratic conquests. State socialism erred seriously by substituting democratic cooperation with egalitarian coercion. All this is now coming to an end. Presently, the material basis of democracy requires taking into account inequality and the progressive end of economic growth as core objectives of the public agenda. We have no modern democratic experience without growth, but we have known democracy and primitive cooperation without growth. That is why it is so important today to understand the evolutionary foundations of democracy and cooperation. The issue is not that democracy is incompatible with degrowth, it is that democracy will only be possible with degrowth and degrowth will only be socially achievable with democracy (Garrido, 2009). The totalitarian dreams of authoritarian communism with zero growth, as proposed by Harich (1978), are not only undesirable, they are also impossible.

History and democracy's biopolitical foundation show us how democracy is solidly anchored in our species, but they also unveil the limits and conditions in which it will be feasible in the not-too-distant future. The brutal asymmetry between cooperation in primitive communities and the possible cooperation in postindustrial societies commends us to flee from any form of neo-primitivism when designing the institutions of the future. A democracy of degrowth and sustainability must be highly cooperative and participative, far from the coercive and elitist market democracy.

4.4.4 DELIBERATIVE DEMOCRACY

In our species, cultural selection is the most frequent and dominant form of natural selection either by individual mechanisms of reflective rationality, or collective cultural processes such as biocultural memory. Therefore, rational and reasonable decisions, according to the well-known distinction of Rawls (1993), play a fundamental role in shaping preferences, judgments, and individual choices, but also, and especially, collective ones. This is the basis of the so-called deliberative democracy, where a decision is not democratic if it is not protected by prior and free deliberation, based on public reasons, among a community of equals: without dialogical and public reasoning there are no democratic decisions. Deliberation is a formal and material prerequisite that has been assumed as a condition of validity in the preparation and parliamentary approval of laws or in the motivation of judicial sentences in the Rule of Law (Ferrajoli, 2010). A law that has not been debated is not valid, or a judgment or judicial order that is not reasonably motivated are invalidated as of right.

Deliberation seems to be inscribed in the very genesis of human collective decisions and cannot be overlooked without bypassing democracy itself, but deliberative democracy must necessarily take into account the interests of those who cannot speak (the silent biotic community) or those who cannot yet speak (the absent community of future generations). This requirement is inscribed in the demands of ecological ethics (Garrido, 2012) and in the factual demands of the environmental crisis. There is indeed no direct relationship between democracy and sustainability, but that does not prevent us from referring to a relationship of possibility and even probability (Bronley, 2016). Of all the forms of collective decision-making, or decisions that affect the commons, deliberative democracy is the only one that can link democracy and sustainability by incorporating the interests of voiceless actors. And in turn, of all the models, deliberative democracy is the most likely to efficiently coordinate democracy and sustainability (Wiromen et al., 2019). This is because it uses decision-making processes based on collective intelligence and public reason, inherent to the dialogical rationality of human language, in which estimation and valuation always exceed strictly individual interests and perspectives and therefore claim to be objectively and universally valid (Habermas, 2010). Therefore, it is necessary to activate mechanisms of altruism to the greatest degree possible through multilevel selection that include the representation of an abstract community of its own (humanity, the biosphere, etc.). This latter responsibility falls on ecological ethics that compels the community of equals to expand (Singer, 2002), as we have defended.

For deliberative democracy to increase the chances of operating as a cooperative and inclusive democracy, it is necessary to reinforce some features of institutional design in deliberative decision-making: the representative dimension of democratic decision-making is inescapable, even at very low scales of complexity. Accepting that only those who have a voice can defend their interests, as in the case of certain direct democratic programs, is incompatible with the principle of feasibility in complex societies (Domenech, 1998) and makes it unviable to include the interests of those without a voice (the absent, future generations, the biotic community, etc.). Decisions cannot be the product of aggregate individual decisions, but of the decision median. The consequences of collective decisions must be the same for all, whatever individual decision has been adopted. An experiment conducted by a team led by J. Nowak demonstrated, through a simple game of public goods, that when individuals make future decisions through a deliberative vote model, the result is favorable to the interests of the future generations and not only to the sum of the private interests of those who took the decisions. The groups that made decisions through an aggregation model of individual votes were blind to the future (Hauser et al., 2014). Likewise, the tragedy of the commons described by Garret Hardin (1968) highlights the ecological tragedy of isolated individual decisions that didn't follow any cooperative norms, with a money-based cost information system (Mayumi & Giampietro, 2006) and without any deliberative framework. Faced with this type of decision-making, a non-cooperative model of the prisoner's dilemma, we can only counter-propose a model of deliberative democracy with the restrictions we alluded to.

In accordance with Wiromen et al. (2019), deliberative democracy is therefore very useful from an agroecological perspective, for institutional design in three specific fields: (i) the shaping of preferences properly adapted to the demands of sustainability; (ii) the public assessment of standards; and (iii) the legitimizing of institutions and policies. Only by making rationally and reasonably based decisions through an open, regulated, continuous and free dialogue process will there be any chance of success in modifying and introducing new preferences compatible with ecological ethics, evaluating environmental standards and legitimizing the necessary political changes to face the evolutionary challenge of sustainability. Any other authoritarian or non-cooperative alternative will be more costly. It will also be less likely to meet with evolutionary success, as indicated by the theoretical framework this work and Political Ecology.

5 Scaling Agroecology

After describing the twentieth century's abrupt environmental changes on a global scale, especially after the Second World War, John McNeill (2000, 4) concluded that humanity was involved in a "gigantic experiment without control". A few years later, Rockström et al. (2009a) showed that the anthropogenic pressures on the terrestrial system had surpassed certain thresholds that guaranteed ecological dynamics on a planetary scale. From an economic standpoint, Piketty (2014) showed that the recent evolution of global economic governance had generated levels of concentration of income and social inequality that were comparable to those of the nineteenth century; furthermore, the trends were clearly worsening. As we saw in Chapter 3, the current crisis is the corollary of all these tendencies, and a singular crisis, in which multiple crises converge, has been shaped (George, 2010), exposing the limits of modern civilization (Garrido Peña et al., 2007; Toledo, 2015).

The notable growth of global instability since 2008, as economic, political, social, environmental, and climatic turbulence has intensified, has been identified by Wallerstein (1974, 2005) as evidence of the terminal crisis of the world-system that emerged around five centuries ago, in the era of European expansionism (Braudel, 1995). Then, the rapidly expanding frontiers of ecological appropriation and social exploitation led to unprecedented levels of accumulation of capital and power, paving the way for the institutional foundations of capitalism. The symptoms of the current crisis suggest that we are witnessing a phenomenon hitherto unknown that will not be solved with the same pattern of response adopted in the cyclical crises of the past, i.e., "by putting nature to work in powerful new ways" through "new technologies and new organizations of power and production" (Moore, 2015, 1).

The magnitude of the transformations needed at this time of "historical bifurcation" may have had only two precedents in the course of human history: the neolithic revolution, with the advent of agriculture 10,000 years ago; and the Industrial Revolution, a process that begun some 300 years ago, and a close source of the current civilizational crisis. These two "revolutionary moments" gave rise to large-scale and far-reaching transformations in humans' integration in the biosphere, as well as in the corresponding societal configurations (González de Molina & Toledo, 2011, 2014). Everything points to the fact that equally radical socioecological transformations will be necessary to deal with both the causes and the effects of the – now inexorable – climate changes, as well as the other symptoms of the global crisis corroding the foundations of modern societies. However, unlike the previous crises, which arose locally and spread globally, the necessary changes in the "age of globalization" (Giddens, 2000) and the "Empire" (Hardt & Negri, 2000) are going to require profound transformations related to the governance of food systems' metabolism at different scales, from local/territorial levels to a global scale.

5.1   THE NATURE OF CHANGE: THE METAMORPHOSIS OF THE FOOD SYSTEM

Stopping and reversing the crisis, building a metabolic regime grounded in sustainability in its different environmental, social, cultural, and economic dimensions is, therefore, a task that cannot be deferred. Agroecology has the responsibility of building sustainable food systems and reducing the corporate food regime (CFR)'s excessive metabolic profile without increasing social or territorial inequality. The role of Political Agroecology is to design and produce institutions that promote the agroecological transition in that direction. The change cannot be sudden, but it cannot either be postponed until today's power relations, that currently favor the continuity of the CFR, begin to change. Nor can the renewal be gradual and cumulative. It must take our existing circumstances as the point of departure to drive meaningful change. Political Agroecology is far removed from the reform/revolution dichotomy, characteristic of modern thought. Rather than portraying it as an agroecological revolution, as evoked by prominent Agroecology theorists (Altieri & Toledo, 2011; Sosa et al., n.d.), we believe the transition is best described as the metamorphosis of technical–institutional arrangements regulating food production, processing, distribution, and consumption patterns.[1]

Using the biological analogy of metamorphosis enables us to emphasize two interdependent aspects, one of a political order (the collective action plan) and another of an intellectual order (the thought plan). From the political perspective, the notion of metamorphosis allows to overcome the dilemma synthesized in the title of the book by Rosa Luxemburg *Reform or Revolution* ([1900] 2010). In the same way that a caterpillar turns into a butterfly, metamorphosis combines the gradual nature of transformations within the system, as advocated by reformists, with the immediate break with the systemic order, as defended by revolutionaries. The intellectual implication is the need to recognize that forces of transformation have no center of gravity. These forces are scattered around the world, organized into structured networks at different scales, from the most hidden corners to incipient initiatives aiming at articulating global society, in our case, around the political agenda of Agroecology and food sovereignty. This means that the processes of change are not guided by a universal theory put into practice by avant-garde forces. Rather, processes of change are already underway, expressed through countless agroecological practices paving the way for a new regime to be built. In this sense, an understanding of change as a metamorphosis is in line with the thesis defended by Holloway (2011). The latter

---

[1] The metaphor of agroecological metamorphosis is inspired by Edgar Morin (2007, 2010). "When a system is unable to deal with its vital problems, it degrades or disintegrates or is capable of eliciting a meta-system to deal with its problems: it metamorphoses" (Morin 2007, 179, 181).

author holds that the hegemonic system needs to be cracked through concrete social experiences that bring about increasing autonomy with regard to the production modes commanded by the logic of capital: "the only way to think about changing the world radically is with a multiplicity of interstitial movements flowing from the particular" (Holloway, 2011, 15).

Agroecology is becoming increasingly consolidated worldwide as a critical theory that radically questions industrial agriculture, while simultaneously providing the conceptual and methodological bases for the development of economically efficient, socially just and ecologically sustainable food systems. As a social practice, Agroecology is expressed in the diversity and creativity of the forms of resistance and struggle found among peasants, in particular their strategies for constructing autonomy from the input and labor markets. As a social movement, Agroecology mobilizes actors practically and theoretically involved in its construction, as well as growing sectors of the population to fight for social justice, collective health, food and nutritional sovereignty and security, a solidarity and ecological economy, gender equity, and more balanced relations between the rural world and the cities. In essence, Agroecology produces a synergy between three forms of understanding, condensing its analytic approach, its operational capacity and its political advocacy into an indivisible whole.

Agroecology has in fact sparked countless, mostly local, sociotechnical innovation experiences over the last decades involving social organizations, researchers, extension agents, cooperation agencies, public managers, and consumers in different parts of the world. The results of these technical, economic, social, and environmental experiences are today internationally recognized as practical expressions of a consistent strategy that help to face global challenges. These challenges have been identified in the "Agenda 2030 – Transforming Our World", which determined 17 Sustainable Development Goals (SDGs) (UN, 2015) and through global initiatives related to the political agenda of sustainable development such as the Paris Agreement on Change (UN, 2015b), the United Nations Decade of Action for Nutrition (UN, 2016) or the Thirteenth Conference of the Parties on Biodiversity. Since the 2008 world food crisis, various United Nations bodies have published important documents confirming the notion that the agroecological approach offers consistent responses to the current accentuation, global spread and mutual interweaving of food, energy, ecological, economic, social, and climatic crises (IAASTD, 2009; De Schutter, 2011; HLPE, 2012; UNCTAD, 2013).

The empirical evidence gathered from these experiences in different regions of the world (Brescia, 2017; Oakland Institute, 2018; Biovision, 2018; IPES-Food, 2018; Mier et al., 2018) also shows that they do not result from a planned, mega blueprint of top-down change (Scott, 1998) aimed at an abrupt and linear transformation of the dominant regime. On the contrary, agroecological experiences are giving rise to non-linear agroecological processes that are complex and adjusted to local socioecological and historical specificities. They represent processes of sociotechnical change driven by organized local collective action (IPES-Food 2016, 2018). These experiences entail innovations in the field of local institutions and the creation of new markets, in addition to technical innovations (Hebinck et al., 2015).

However, one of the main lessons of the history of agriculture is that the super-seding of one pattern of technical and economic organization of agroecosystems by another has never occurred as an automatic outcome of new technological dis-coveries. The large-scale adoption of technical innovations tends to run into strong political-institutional and cultural obstacles, even when these new technologies have already shown they could provide solutions to some of the deep dilemmas faced by societies (Thirsk, 1997; Mazoyer & Roudart, 2010). This is why Agroecology remains limited to niches of social innovation, posing no threat to the institutional foundations that sustain the CFR. This is despite the fact, as presented earlier, that the CFR is increasingly linked to the risks of socioecological collapse now evident within the Anthropocene era, particularly since the second half of the twentieth cen-tury, the period known as the *Great Acceleration* (Costanza et al., 2007).

Thus, agroecological transitions cannot be represented as binary, unidirec-tional, and deterministic processes of change. In many cases, the transition paths are obstructed by unforeseen hostile situations, or even suffer unexpected setbacks before moving on, spurred by the coordinated action of local actors. Converting these experiences into cracks in the system and into the drivers of the agroecological meta-morphosis is not an easy task. Nor will it take place through a simple accumulation of experiences. It is necessary to direct and coordinate such experiences so that, gathered under a common strategy, they may unfold all their metamorphic potential. It is the responsibility of Political Agroecology to draw up a strategy that organizes the different levels of collective agroecological action to make this possible.

5.2   A STRATEGY FOR CHANGE: SCALING UP AGROECOLOGY

The turbulent landscape in which we live creates exceptionally favorable conditions for the regime to be contested and influenced by the proposals emerging from agroecological experiences. The global systemic crisis opens "windows of oppor-tunity" for these emerging practices and ideas to transform the regime. "Window of opportunity" was precisely the term used by the FAO Director-General José Graziano da Silva at the opening of the First International Symposium on Agroecology for Food and Nutrition Security in 2014, an activity carried out in the International Year of Family Farming. The time is ripe for the experiences to gain ground, the time has come for the scaling out and scaling up of Agroecology (Gonsalves, 2001; Holt-Giménez, 2001; González de Molina, 2013; Rosset, 2013; Levidow et al, 2014; Parmentier, 2014; Gliessman, 2018), it is time for the cracks to turn into the fabric of the food system, time to make a leap forward in the metamorphosis. The moment has come to fight for the number of agroecological experiences (scaling out) to grow, but above all, it is time to scale up and create a new institutional set-up inspired by the principles of institutional design and the cognitive frameworks described in the previous chapter.

A growing number of authors have been documenting and conceptualizing the phasing processes of Agroecology as a strategy to promote more sustainable food systems. This goal gained great international visibility when the *Second FAO International Symposium on Agroecology: Scaling up Agroecology to Achieve the Sustainable Development Goals* was held in April 2018 (FAO, 2018). However, this growing acceptance of Agroecology at institutional levels has led several authors in the agroecological field to draw our attention to the risks of cooptation by the regime (González de Molina, 2013; Méndez et al., 2013; Gliessman, 2014; Levidow et al., 2014; Giraldo & Rosset, 2016). The concern is systematically fueled by frequent official demonstrations that reduce Agroecology to a scientific approach aimed at developing knowledge and technologies to make modern agriculture more sustainable.

Political Agroecology provides theoretical and methodological references to support institutional innovation processes allowing to apply an agroecological view of sociotechnical change to ever greater scales. Meanwhile, it must ensure that the field is not captured by the dominant food regime to which it is subordinated and relegated as a complementary sector, nor that it becomes encapsulated or systemically rejected. Designing appropriate strategies to put the principles and fundamental values of the agroecological paradigm into practice in hugely diverse and particular social, ecological, and political contexts leads to polarizing debates around Agroecology's phasing processes. A basic principle emerges from these debates: due to the site-specific nature of agroecological practices, transition processes should preferably be driven from the bottom up. In this sense, the processes of agroecological transition are in direct contrast with the intellectual, technical–economic and political foundations of agricultural modernization, a project actively promoted in the "developed" countries after the Second World War, and then spread to the Third World through the so-called Green Revolution.

In theoretical terms, agroecological metamorphosis points to the need for a paradigmatic break with structuralist models of development that conceive sociotechnical changes as the result of interventions by the State and/or of external economic agents, following a predetermined and guided path, based on stages of development or through a succession of dominant modes of production. Despite the ideological differences between the theory of modernization and the neo-Marxist theory, both have paradigmatic similarities in so far as they advance deterministic, linear and externalist visions of technical and social change (Long & Ploeg, 1994). Both theoretical–political projects idealized and promoted massive processes of depeasantization. On the Soviet side, the communist regime imposed extensive collectivization of peasant and landowner expropriated lands. On the other side, "the end of the peasantry" had to take place through agricultural modernization, understood as the conversion of peasants into agricultural entrepreneurs.

Having received broad political, ideological, financial, and in many cases military support from State governments, the modernization project spread territorially, accelerating the depeasantization in countries of the North and South. For Hobsbawm (1994, 288–299) the most dramatic change in the second half of this century was the death of the peasantry. Echoing the English historian, a wide range of academic and political circles has predicted the end of the peasantry (Mendras, 1967) as an inevitable destiny in the face of capitalism's encroachment into the rural world. Empirical reality, however, shows that global agrarian history does not follow the destinies theoretically envisaged on the basis of the modernization paradigm. Rather than a single script of rural development determined by the forces of globalized and globalizing markets, observable the world over, we have diverse trajectories of development of agroecosystems, influenced by locally written scripts and energized by the multiple creative forms of resistance and struggles for the emancipation of peasant farmers. A central element in this resistance is the continuous construction, improvement, amplification, and defense of a self-controlled base of local resources, composed through the mutual interconnections between nature's goods and institutional arrangements of social integration.

5.3 PEASANT AGRICULTURE: THE COCOONS OF AGROECOLOGICAL METAMORPHOSIS

It is precisely these experiences of peasantization and repeasantization of agriculture, currently taking place on a global scale, which underpin the scaling-up process. In fact, agroecological transition trajectories can be assimilated to these processes which in turn are countermoves through the "de-commodification of food systems" (Petersen, 2018). The main strategic objective is to remove peasants and farmers from the CFR. Unsurprisingly, family farming remains the main way of organizing world food production. Family farmers are still operating 75% of the world's agricultural land and are likely to be responsible for the majority of the world's food and agricultural production as we saw in Chapter 3 (Lowder et al., 2016). They possess high agroecological potential. Among other reasons, they can be closer to rural rationality and practices that make it possible to sustainably manage agroecosystems (Altieri & Toledo, 2011). Traditional peasant communities have also developed highly efficient methods of cooperative and communal management of resources (Ostrom, 1990, 2001), which are of great interest to Agroecology. As a response to the industrialization of the food system, *repeasantization* has been taking place for some time, accentuated by the CFR crisis. This process is characterized by the struggle for autonomy and decommoditization and by the quest for local solutions to global problems. It goes much further than a mere resistance movement (Ploeg, 2008). This movement is becoming ever more widespread, especially in the countries of industrialized agriculture.

The characteristics of peasant production have been supporting markets' relatively autonomous local economies, helping to explain their historical continuity in an increasingly hostile world. Analyzing them sheds light on important facets of agrarian reality that remain invisible to dominant structuralist approaches and significantly contributes to elaborating a theory of the agroecological transition (or Political Agroecology). Although peasant agriculture is a multimillennial institution, which is permanently reinvented by a large part of humanity, the dominant scientific paradigm, with its positivist postulate and its reductionist and mechanistic methodological approaches, has been unable to capture the essence of its modus operandi. Behind its theoretically and institutionally constructed invisibility lies a concrete reality that refuses to go away thanks to its capacity for resistance and struggle for emancipation. These intrinsic capabilities surpass the specificities of time and space. They manifested themselves in the agrarian societies of the remote past and in the emergence period of mercantilist capitalism. They are also visible at this very point in time, as the dominant regime is in full swing, both in the Global North and South. Academic and institutional mainstream face this uncomfortable truth, as Shanin

(1966, 5) wrote: "Day by day the peasants make the economists sigh, the politicians sweat, and the strategists swear, defeating their plans and prophecies all over the world".

In dialogue with critical theories of economics (Political Economy, Feminist Economics, Ecological Economics, Neo-Institutional Economics) and inspired by the Chayanovian approach to the analysis of the peasant economy, the analysis of peasant production from an agroecological viewpoint sheds light on the "hidden part of the iceberg", or on the non-mercantile economic circuits through which peasant agriculture mobilizes a large part of its means of reproduction (see below). In addition to securing greater levels of resilience in the face of an increasingly turbulent and unpredictable world, peasant agriculture is the vehicle of consistent responses to the socioecological dilemmas faced by contemporary societies that have been provoked by the decisive contribution of agribusiness, that is, by a form of agriculture integrally governed by "market logic".

The Russian economist Alexander Chayanov was a prominent author who was opposed to the anti-peasant consensus imposed in his country with the Bolshevik revolution at the beginning of the twentieth century. Describing a set of principles governing the economic functioning of peasant units and differentiating them from the capitalist mode of production (Chayanov, 1966a [1928]), he explained why they are not directly governed by market rules, although they are conditioned and influenced by the capitalist context in which they operate.

The essential aspect that distinguishes peasant economic organization from its institutional environment is that the labor force that drives the capital involved in the production unit is the family itself. This means that peasant agriculture is not organized to extract and appropriate the wealth generated by the work of others, that is, by the generation of surplus value. In addition, while being both the owners of the means of production and its workers, the peasantry depends on the conservation – and, whenever possible, on the expansion – of the productive heritage, implying a specific resource management rationale. This technical–economic rationale ensures relative autonomy from the markets and cannot be explained exclusively by the same factors that determine the functioning of capitalist units, that is, the market, the technological means, the availability of land, etc.

Chayanov convincingly demonstrated that peasant production units are structured according to the strategic deliberations of the families themselves, conducted in the course of their life cycles. "We will fully understand the basis and nature of the peasant farm only when in our constructs we turn it from an object of observation to a subject creating its own existence, and attempt to make clear to ourselves the internal considerations and causes by which it forms its organizational production plan and carries it into effect" (Chayanov, 1966b [1925], 118).

Although they have been forgotten for several decades, Chayanov's theoretical contributions remain of prime importance in the contemporary debate on agricultural sustainability. Using a Chayanovian approach, Ploeg (2010, 2013, 2018b) proposed an interpretation of the objective conditions of the peasantry's historical presence at the beginning of the twenty-first century, which is very useful for Agroecology. "Central to the Chayanovian approach is the observation that although the peasant unit of production is conditioned and affected by the capitalist context in which it is operating, it is not directly governed by it. Instead, it is governed through a set of balances. These balances link the peasant unit, its operation and its development to the wider capitalist context but in complex and definitively distinctive ways. These balances are ordering principles" (Ploeg, 2013, 5).

From this analytical perspective, peasant production units can be understood as the sociomaterial expression of a strategy of social reproduction actively built over time, depending on internal variables (such as family life cycle, availability of access to goods, etc.) and external ones (participation in local organizations, specific links to markets, access to public policies, etc.). At the same time, agroecosystems can be regarded as an expression of strategic projects led by the peasant nuclei (be they families or rural communities), closely interacting with the dynamics of ecosystems and the political–institutional conditions of the environment.

According to Scott (1976), the peasantry's ability to "re-exist" in a world that is systematically hostile to its historical perseverance is explained by their adopted logic of labor organization, defined as a "moral economy".[2] By integrating values, norms, collective memories, beliefs, and shared experiences, peasant economic rationale is strongly influenced by the cultural references that condition ways of perceiving, interpreting and acting within the reality in which peasants live and produce. From this point of view, it is radically different from the logic of corporate *Homo oeconomicus* motivated by self-interest and maximizing opportunities.

Two components lie at the heart of this moral economy: a logic of work organization that prioritizes subsistence, and control over means of production (Bernstein, 2001). In this sense, peasant economic rationale organically follows an ecological rationale (Toledo, 1990) in the management of nature's elements that are part of their means of production (their objects of labor). Although increasing the land's physical yields is an important objective of peasant moral economy, it is only one among several others that are taken into account in the planning and evaluation of the labor process. These are continuously defined taking into account the ecological, demographic, cultural, institutional, economic and political conditions in which the peasant nuclei operate, which corresponds precisely to the sense of balance identified by Chayanov, leading him to conclude that peasant agriculture should be understood as an art.

Moved by an economic rationale in stark contrast with the business and capitalist logic of agroecosystem management, peasants who are major resistance players structure their labor processes so as to avoid relying structurally on the mercantile relations of the dominant sociotechnical regime. To do this, they systematically act to defend and expand a self-controlled resource base, where they mobilize the factors of production for the labor process. In addition to labor objects directly appropriated from nature (animals, seeds, soil, water, etc.), the self-controlled resource base is composed of social resources (knowledge associated with agricultural work, social networks of reciprocity, etc.) and work tools (machines, equipment, infrastructures). Self-controlled social and material resources are integrated into an organic and indivisible unit that does not allow for an analytical differentiation between "labor and

---

[2] For Scott (1976), the concept of peasants' moral economy is based on three basic principles: safety first (avoiding risks); the ethics of subsistence; and justice associated with reciprocity.

capital", "economic production and ecological reproduction", and "manual labor and intellectual work".

This form of organization of agricultural production has a distinctive characteristic: a large part of the resources employed are not mobilized in the productive process as a commodity (as capital). This means that it is viable to produce using the resources reproduced in previous production cycles, thus highlighting the central role of work and information in agroecosystem management. Therefore, the technical efficiency of converting resources into products (the intensity level) depends essentially on the quantity and quality of both the work and experiential knowledge (flows of information), and not on the acquisition of factors of production and external knowledge.

This way of structuring the labor process is directed toward the creation of added value, whether converted into money or not. This "autonomous" (or "organic") growth of value added at the scale of the production unit is driven by investment in labor and knowledge over the course of the various productive cycles. "This occurs through a slow but persistent growth of the resource base, or through an improvement of the 'technical efficiency'. Mostly, however, both movements will be combined and intertwined and will thus obtain an autonomous moment of self-enforcement" (Ploeg, 2010).

Faced with hostile institutional environments, labor-based intensification paths are precisely the means by which, farm families, at the level of production units, combine their resistance with their struggle for autonomy and sustainability. The following five processes are decisive in the causality chain of resistance, autonomy, intensification and sustainability (Ploeg, 1993, 2007, 2015; Altieri et al., 2012, Egea-Fernández & Egea-Sánchez, 2012; Tittonell et al., 2016). The first process is that agricultural work organization is attuned with coproduction dynamics, that is, based on synergistic interactions and mutual transformation between human work and the work of the rest of nature. This agroecosystem management logic helps management practices to converge with the ecological dynamics of ecosystems, shaping an organic unity between economic production and ecological reproduction. Second, the valorization of ecological capital in the labor process is the means by which dependence on commercial inputs is reduced. This ensures better conditions for resistance in economic environments where production costs systematically increase, while output prices vary erratically due to the deregulation of agricultural markets. Third, natural goods are conceived as an integral part of peasant families' patrimony and not as commodities that can be valued in the market. This marked contrast with the capitalist logic of natural asset appropriation leads to conservationist agricultural practices. Fourth, the management of complex and biodiversified agroecosystems is based on scope economies, that is, those that seek to reduce total costs using the synergies between various productive activities coordinated by means of a single management process. This management style relies on the circularity of economic–ecological flows at the rural territory level. It thus reproduces a basic principle in the functioning of natural systems: the residues of one species are used as food for another or become necessary elements for the reproduction of ecological processes at the scale of the landscape (Guzmán Casado & González de Molina, 2017). Fifth, the productive diversification of agroecosystems plays an essential role in farm families' food supply. Consequently, a large part of the food is secured without the need for mercantile exchanges.

However, Chayanov's theoretical elaborations left unexplored a key aspect of the economic functioning of peasant management agroecosystems: the specificity and economic value of women's work. Subsequent contributions from Feminist Economics have been decisive in understanding that units of production in family farming work as hubs of cooperation and conflict according to gender inequalities, fed by deep-rooted patriarchal cultures systematically found in the peasantry. As we will see later, overcoming gender inequalities in production units, communities and peasant organizations is decisive for the deployment of larger-scale agroecological transition paths.

On the other hand, it is on the microscale of peasant production units that market trends, technical requirements, public policies, climate change, and other macro-structural influences are interpreted and translated into actions in practice, according to the strategic coherence of peasant families and communities. "The peasant unit of production is precisely the institutional form that distantiates farming in a specific, strategically ordered way from the (input-) markets, whilst it simultaneously links it (also in a specific, strategically ordered way) to other markets (on the output side of farming)" (Ploeg, 2010).

This ability to translate the signals emitted from the macroscale into a specific strategic course of action defined by peasant families (or communities) at the microscale means the peasant management agroecosystem operates like a rhizome – in line with the metaphor developed by Deleuze and Guattari (1995). The peasant agroecosystem is not subject to universal and totalizing itineraries, it has multiple horizons and can follow different directions; its trajectory does not follow any straight lines defined by Cartesian and binary calculations, it is open to experiment. The agroecosystem connects with other rhizomes building new paths. Connections are diversified, forming complex networks that can be extended and disseminated, creating new sociomaterial realities. The rhizomatic movement operates underground, countering the regime's visible movements. The rhizome, used as a metaphor for the peasant farm, symbolizes the peasantry's ethical–aesthetic–political resistance.

In short, although much of these peasant resistances may appear irrelevant when analyzed in isolation, taken together their practices point to consistent ways of building local solutions to serious global dilemmas caused by modern food systems. These practices reproduce relatively autonomous and sustainable agrarian metabolisms, shaped by technical–institutional arrangements that organize agricultural labor in accordance with fundamental elements also present in the organization of the work of nature: diversity; the cyclical nature of processes; adaptive flexibility; interdependence; and ties of reciprocity and cooperation.

5.4   COUNTERMOVEMENTS FAVORING THE DE-COMMODIFICATION OF FOOD SYSTEMS

The market, as the main vector for inducing industrial metabolism in food systems (González de Molina & Toledo, 2011), disrupts the organic unity of economic production and ecological reproduction, responsible for the multimillennial evolution of agriculture, generating increasingly entropic and globalized metabolic profiles. Under the aegis of the neoliberal project, the CFR has led to rapid restructuring of agricultural markets, followed by the accelerated disarticulation of national, regional and local production and food supply systems (Lee & Marsden, 2009). In addition to promoting intrinsically unsustainable metabolic patterns (Krausmann & Langthaler, 2019), this process of "uprooting" agricultural economies transfers important plots of power over the governance of food systems to a small number of economic actors freely operating in global markets, and exclusively driven by profit maximization (Ploeg, 2018b).

Political Agroecology proposes to restructure dominant economic processes in the CFR reestablishing circularity between agriculture and nature (Jones et al., 2011). This implies restoring the power of governance over the processes that link the production, processing, distribution, and consumption of food by the actors directly involved in those activities (Lamine et al., 2012). In other words, it is about reconstituting democracy in food systems (Renting et al., 2012; Pimbert, 2018). And that is precisely how Agroecology understands food sovereignty demands. Consequently, a critique of the power exercised by corporations in the agroindustrial and financial sectors in shaping the institutional arrangements that regulate food production, distribution and consumption is central to Political Agroecology. Over the last decade, there have been abundant academic studies in the field of economic sociology devoted to alternative food networks, an emerging social phenomenon identified in various regions around the world that represent local responses to the negative impacts of globalization and the corporate concentration of food markets (Wiskerke, 2009; Brunori et al., 2012; Lamine et al., 2012; Ploeg, 2012; Perez-Cassarino, 2013; Niederle, 2014; Hebinck et al., 2015; López García et al., 2015; Valle Rivera & Martínez, 2017).

The link with these local production and food supply networks lies in the fact that they are not integrated (at least not entirely) into the CFR. They thus deviate from the hegemonic sociotechnical script and are materialized by means of lines of innovation in accordance with the perspectives, values and objectives negotiated by actors organized in territorially referenced networks. Taken together, these initiatives to relocate food systems can also be interpreted as a "countermove" (Polanyi, 2001 [1944]) to commodify trajectories. In his classic work, *The Great Transformation*, Polanyi analyzed the institutional changes that shaped modern capitalism and emphasized the importance of social countermoves that opposed the imposition of "fictitious goods", i.e. goods and services that were not produced to be marketed, such as land and labor. According to the author, "the commodity fiction disregarded the fact that leaving the fate of soil and people to the market would be tantamount to annihilating them. Accordingly, the countermove consisted in checking the action of the market in respect to the factors of production, labor, and land" (Polanyi, 2001, 137).

The essence of these countermoves would therefore lie in their struggles against the commodification of a growing number of plots in the social and natural world. In this case, commodification would be countered by a moral order imposed to protect the human fabric, nature and the organization of economic processes themselves (Niederle, 2014). The key point in Polanyi's analysis is the fact that the economic functioning of human communities depends on the presence of well-established institutional structures that combine, to varying degrees, three main forms of social integration: reciprocity, redistribution and trade exchanges (Polanyi, 2012).[3] His theoretical insights and ontology are highly useful to examine the current "governance pattern" imposed by the dominant food regime (Schneider & Escher, 2011). Two Polanyian contributions are particularly relevant for Political Agroecology. The first is the realization that, in a complex society, farmers' economic behavior is strongly determined by the institutional environment and the social relations they are part of. Forms of social integration are institutionalized through the socialization of practices based on collective action mechanisms and rule and value systems. In this sense, Long (1986) describes the commodification of agriculture as a process of "institutional incorporation". Hence, innate psychological dispositions of behavior, such as those proper to neoclassical *Homo oeconomicus*, are not decisive in shaping the market economy, as advocated by Hayek (1944), a prominent theorist of economic liberalism and a contemporary of Polanyi.

The second contribution is the fact that the deepening of capitalist sociability and mercantile exchange in food systems triggered a number of "countermoves" regarding commodification. From the point of view of agriculture, resistances to the dominant sociotechnical regime are materialized through localized labor organization experiences that reflect disputes over the control of productive resources (land, water, biodiversity) and of the agriculture markets themselves. In these countermoves, elements of nature mobilized by agricultural work are not conceived and managed according to the commodity rationale. Agricultural markets, in turn, are understood as social constructs or dispute arenas, and not as abstract economic systems supposedly self-regulated by "invisible hands", as postulated by liberal theorists.

To summarize, Polanyi's reflections inspire Political Agroecology insofar as he characterized "the historical construction of the market economy as an immense and violent artificial social process, which did not conform to human nature's presumed characteristics, but rather to an ideological, axiological and political commitment that is radically different from the previous forms in which human groups had organized and integrated material resources and their sustenance. According to his theoretical critique, the market economy is a disintegrator of society's human essence, necessarily implying the need for political action that is transformative and regulates the market, articulating his reflections as thought for action" (Sánchez, 1999, 1).

As "thought for action", Political Agroecology is both a political theory of the socioecological crisis of food systems and a socioecological theory to design political institutions that regulate sustainable agricultural metabolisms. Through fruitful interactions with critical perspectives from the social sciences, this theoretical aspect of Agroecology opens a vast field to develop "valuation languages". These languages overcome the limitations of economic productivism, a hegemonic cognitive framework in the public arena in which policies for agriculture and food are defined and evaluated. For this, Political Agroecology is a disciplinary approach that draws up

---

[3] For Polanyi, the economic process takes place on two differentiated, yet interconnected levels. The first is the interactions between humans and their environment; the second refers to the institutionalization of these interactions.

a radical critique of liberal ideology and the institutional foundation of neoclassical economics, that is, the capitalist market (Garrido Peña, 2012).

It is thus Political Agroecology's function to support the strengthening of countermoves of resistance and reaction to agriculture's commodification. Although inconspicuous, these countermovements led by farmers and rural communities, and also by urban social groups, especially consumers, are widespread throughout the world. In contrast to conventional agricultural development trajectories, centered on increasing degrees of commodification of agroecosystems through the intensive use of commercial factors of production and the continuous scaling up of commodity production, these alternative trajectories are characterized by the growing relevance of economic–ecological transactions governed by relations of reciprocity. In order to build and/or maintain "institutionalized distantiation from markets" (Ploeg, 1990), reciprocity mechanisms are developed by strengthening local cooperation and collective action mechanisms as well as developing co-production dynamics, understood as relations of reciprocity between human beings and living nature.

By revaluing endogenous (natural and social) resources (Oostindie et al., 2008) and developing local mechanisms for social regulation of economic and ecological flows, these socially and ecologically contextualized initiatives reorganize patterns of food production, distribution, and consumption. They therefore configure sociotechnical innovation experiences in tune with the agroecological paradigm, because they combine high levels of economic efficiency (intensity) with ecological sustainability. In that sense, they reveal themselves as powerful expressions of emerging processes to find local solutions to challenges that also manifest themselves globally. In other words, they can be understood as social forces driving the agroecological metamorphosis (Morin, 2010; Petersen, 2011) in the cracks of the dominant food regime (Holloway, 2011). These niches of sociotechnical innovation are the cocoons of the agroecological metamorphosis. They have several dimensions and levels of complexity. They cover several geographic scales and articulate sociotechnical networks that mobilize the agency of multiple social, individual and collective actors. Emerging social forces incubate in these rhizomatic networks shaped by sociopolitical connections and economic–ecological flows.

5.5 SCALING UP TERRITORIES

All these experiences have emerged in socioecological contexts where the adoption of a territorial approach perspective has been central. This approach is key to the cooperative coordination of the phasing of Agroecology. In fact, one of the recurring features of stepping-stone initiatives is that they are driven by locally organized agents (Ipes-Food 2016, 2018; Mier et al., 2018). Indeed, in the historical context of neoliberal globalization and the CFR's domination, territories and localities acquire a specific meaning as opposed to the global governance of agrarian metabolism. In accordance with the previous chapter, adopting a territorial perspective implies developing a "polycentric system of governance" (Ostrom, 2015b) with the capacity to rebalance the power relations between the State, the market, and civil society. The territory is a privileged locus where the agroecological approach has been applied for the sociomaterial transformation of food production and supply systems. It represents a new geopolitical and geoeconomic standpoint to design institutional arrangements suitable for Agroecology on a larger scale.

It is within a territory that ecological goods, economic activities, local (individual and collective) actors and their cultural repertoires are coherently combined based on locally negotiated and defined strategic perspectives and projects. It is within territories that new institutions are built to mobilize (material and immaterial) resources characteristic of "territorial capital" (Ventura et al., 2008) as common goods (Ostrom, 2015a). In this sense, common goods feed the "expolary economies" (Shanin, 1988) and processes of endogenous development (Oostindie et al., 2008), providing an escape from the imprisoning bipolarity of the State's rules and the capitalist market. Territories represent, therefore, a decisive scale to restore the democratic governance processes of food systems (IPES-Food, 2016).

Markets built by territorialized sociotechnical networks, also defined as nested markets (Hebinck et al., 2015) or territorial markets (CSM, 2018), are institutions that link specific producers and products to specific consumers through specific channels, according to specific rules and values. Therefore, they do not lend themselves to being coordinated by general conventional market mechanisms. This capacity to build distinction (in Bourdieu's sense, 1979) regarding the products circulating in conventional food markets (in terms of quality, technical process, and social origin of production, price, availability, etc.) is decisive to shape and defend socially regulated markets in the territorial sphere against attempts of appropriation by private groups (Ploeg, 2015).

Knowledge is another key resource also handled as a common good in territorial-based agroecological innovation networks. As a result of a territorially contextualized social construction, knowledge circulates freely in the network (Morgan, 2011). Through this network circulation, local (traditional) contextual knowledge resulting from the historical processes of perfecting co-production is mobilized and recombined with scientific–academic knowledge to feed learning processes based on local experimentation, whether technical or socio-organizational. In this way, the values and principles of Agroecology are materialized in social practices adjusted to specific territorial contexts, reflecting the location's social, technical, political, and biocultural contingencies (Francis et al., 2003; Mendez et al., 2016).

Two interrelated characteristics define this pattern of economic organization on the scales both of "strategic distance from markets" and "relatively autonomous and historically guaranteed socioecological reproduction" (Ploeg, 1993). Both are diametrically opposite the conventional trajectories of agricultural development according to the paradigm of modernization. They are "economies of opposition" (Pahnke, 2015). Their structure does not aim at reproducing capital and transferring it to extraterritorial control centers, but to growing and distributing added value among the different players in the network. Instead of the negative environmental externalities generated by the productivist bias of agriculture, the pattern of economic organization stimulates multifunctional agriculture, reversing the mutually destructive relations between economy and ecology. In doing so, it also contributes to increasing socioecological resilience, reducing food systems' vulnerability to the volatility of international markets and increasing climate unpredictability. In this sense, the networks reproduce economies at a regional scale that share the same meaning as peasant strategies of agroecosystem management at a microscale.

Due to the arguments put forward, the territory presents itself as a decisive arena for the aggregation of social forces in defense of food sovereignty, environmental and social justice, collective health, ecological sustainability, peasants' rights and, to mediate all these goals, of deliberative democracy (Petersen, 2011). The territory, in its role of geographic space where power is exercised against the CFR, is also where State policies interact with the sociotechnical network. In this sense, it is the meeting point between the top-down impact of government initiatives and the democratic expression of the needs and demands of active citizenship.

5.6 LOCK-INS AND SYSTEMIC REJECTION

The multiplication and scaling up of agroecological experiences is both an outcome and a condition for the regime's transformation. The idea is to generate a virtuous circle fed by the mutual transformations between the experiences as niches of sociotechnical innovation and the food system. However, a straightforward multiplication of sociotechnical innovation initiatives is insufficient to boost structural changes (Moors et al., 2004; Ploeg et al., 2004; Charão Marques et al., 2012). As Geels (2002) clarified, changes in the regime (mesolevel) depend on complex social and political processes interfaced with transformation dynamics in the niches (microlevel) and the sociotechnical landscape (macrolevel).

Moreover, agroecological experiences cause changes in the regime, but the opposite may also be true. The "path dependence" in the processes of sociotechnical innovation imposed by the CFR, where economies of scale, rationalization, and productive specialization mutually reinforce each other, lock in the transformative potential of innovations generated through experiences (Horlings & Marsden, 2011). According to IPES-Food (2016), in addition to the "path dependence", "these 'lock-ins' include the export orientation of food and farming systems in many countries, based around large-scale monocultures; the societal expectation of cheap food, requiring low-cost (and high externality) commodity production; the compartmentalized and short-term thinking that prevails in politics, research and business, driving short-term, productivist approaches; the 'feed the world' narratives that focus attention on increasing production volumes of staple crops above all else; and the correspondingly narrow measures of success used to identify progress in food systems. All of these lock-ins are underpinned by the ever-increasing concentration of power in food systems, whereby value accrues to a limited number of actors, strengthening their economic and political dominance, and thus their ability to influence the policies and incentives guiding those systems" (IPES-Food, 2018, 2).

A series of recent studies has explored the potential of agroecological innovation niches to overcome the systemic blockages of the dominant regime with the aim of democratizing agrifood systems (Tittonel et al., 2016; Laforge et al., 2017; Petersen, 2017). A recurring finding in these studies is that conventional public policies do not directly generate niches of agroecological innovation. This means that the State has the power to block Agroecology's very scaling-out dynamics. Government policies integrate a complex system of territorial governance that creates or closes spaces (financial, political, ideological, etc.) to deploy Agroecology innovation and diffusion networks. Therefore, they often create unfavorable conditions for socio-technical innovation experiences. Laforge et al. (2017) analyzed different patterns of interaction between public policies and agroecological transition dynamics driven by social innovation niches. Two of them, the patterns of "containment" and "cooptation", undermined their development and diffusion, strengthening the hegemonic regime.

Mechanisms of containment act by creating invisibility and/or marginalizing (often criminalizing) the generated innovations, while the regime's structural elements are materially and ideologically enhanced. Official rules and regulations for the organization of food markets repeatedly illustrate how scaling-out processes are contained. These rules are established to enable large flows of standardized products into retail chains, but these standards seriously undermine the access of diversified agroecological products to markets.

The failure to recognize biocultural knowledge excludes the actors in sociotechnical Agroecology networks from the official innovation systems. Economic productivism criteria adopted to guide research and extension services marginalize the multicriteria approaches of agroecological innovation processes. Financing policies condition access to public resources to the adoption of technological packages generated under the scheme. The latter mechanism is responsible for the progressive abandoning of agroecological practices, such as the use of native seeds, and, at the same time, the creation of farmers' structural dependence on input markets.

Through mechanisms of cooptation, the innovations generated in the niches are partially assimilated by the regime according to its dynamics, values and norms. According to Sherwood et al. (2012), this reflects a "scaling up in name but not in meaning", where many agroecological principles are marginalized in favor of the fundamentals of the status quo. For Smith and Raven (2012), cooptation is a mechanism leading to "fit-and-conform" radical innovations in the regime. This happens through pressures (by way of various incentives) on innovation niches to align with the regime's operating grammar without affecting its internal dynamics.

A classic example of cooptation is the global evolution of organic agriculture. In many cases, the practices of organic agriculture, arising from dissenting attitudes of producers and consumers, brought about transformations in some components of the CFR without compromising its internal coherence (Smith, 2007). The debate on the standardization of organic agriculture or *conventionalization* (Guthman, 2004; Darnhofer et al., 2010; Niederle & Almeida, 2013; Ramos et al., 2018) accurately reflects this regime's ability to selectively incorporate innovations that emerged as responses to the dominant sociotechnical order. We will return to this subject in the next chapter.

Although emblematic cases of scaling out are relevant as sources of inspiration for other territorial contexts, the multiplication of such exceptional experiences remains heavily obstructed by the sociotechnical arrangements established by the CFR. For these scaling experiences to cease to be exceptional and become generalized, it is essential that the practice, theory, and policy of the CFR be overcome. This requires implementing scaling-up processes, that is, vertical climbing and institutional change.

5.7   PATRIARCHY AS A POLITICAL–CULTURAL OBSTACLE TO AGROECOLOGY

In addition to the blockages generated by the CFR, gender inequality also represents a critical barrier to agroecological transition processes. Women are responsible for a large part of the peasant economy, playing a central role in the management of internal economic–ecological flows of agroecosystems, especially those oriented toward the consumption of families. In this sense, women play a crucial role regarding the maintenance of economically and biologically diversified agroecosystems. In addition, they assume most of the workload in the sphere of domestic work and care, two core areas of peasant family economic reproduction. As Feminist Economics has demonstrated (Orozco, 2004), women's work is central in shaping the "invisible part of the iceberg" assumed by peasant economies, ensuring their long-term stability.

Patriarchal cultures, which are widespread in the traditional peasantry, are responsible for the invisibility of peasant women's key role. Inequality in gender relations restricts women's access to land and other productive resources, limits their participation in decisions related to agroecosystem management, restricts their access to the income produced by the family nucleus, and even, recurringly, to the best foods. At the community level, this inequality is expressed through lower participation in social organizations and markets, less access to public policies, formal education, etc. (Silliprandi, 2015; Galvão Freire, 2018).

The processes of agricultural modernization have not reversed this political–cultural picture. On the contrary, traditional gender inequalities have tended to be aggravated by the progressive adoption of entrepreneurial rationality in the organization of agroecosystems. As a consequence, destructuring processes of economic activities dedicated to the family's biological reproduction (production for self-consumption) and technical system (own production of productive resources) are common. Local cooperation mechanisms between women are equally disarticulated by attributing strategies for the technical and social reproduction of agroecosystems to markets. The question has been that of selective modernization, oriented exclusively toward updating technical–institutional mechanisms to increase the level of exploitation of human work and appropriation of nature's goods. Within this "modernizing" process, peasant women have undoubtedly suffered the most.

Documented Agroecology scaling experiences all emphasize that the active participation of women in decision-making at the farm and community levels is both a prerequisite and an outcome of agroecological innovation (Lopes & Jomalinas, 2011). This implies that Agroecology is the best approach to empower women in the food sector, while agroecological practices are strengthened and developed with the political and economic emancipation of women. However, that dual relationship does not arise automatically. Specific policy and methodological strategies are needed to address culturally constructed gender inequalities. To that end, adopting measures to promote women's access to knowledge, productive resources and participation in decision-making within their families and communities are essential, breaking the vicious circles that exclude them from the processes and the benefits of agroecological innovation. In sum, it is necessary that Agroecology, in its role of democratizing food systems, be conceived bottom-up, that is, by food-producing families. This also includes incorporating methodological approaches that are sensitive to generational relationships and other culturally defined processes of social inclusion/exclusion (such as race, religion, nationality, etc.) as they constitute notable obstructions to the development of Agroecology.

5.8 AGROECOLOGICAL-ORIENTED LOCAL FOOD SYSTEMS

The challenge, therefore, is that of upscaling Agroecology, overcoming the institutional blockade and creating a new, much more favorable alternative institutional framework. A key role must be attributed to the cooperation between the different links in the chain, in such a way as to break with the isolation and fragmentation of experiences. To achieve this, various instruments of social mobilization and innovation already available in the agroecological movement itself and even in co-produced public policies must be brought together to create food systems at a territorial or local level which, on the one hand, operate a leap of scale of agroecological experiences and, on the other, expand the areas of resistance to the CFR, challenging its hegemony. We propose, therefore, as the main strategic objective or as a road map of the agroecological movement the construction of *Agroecological-oriented Local Food Systems* (ALOFOODs hereafter). What do these systems consist of? In the creation and consolidation of a new food regime, an alternative to the dominant one, covering the largest possible ground, gaining hegemony with respect to the CFR and supported by both the power of social movements and their socioeconomic viability, thereby generating broad areas of food sovereignty and sustainable production, i.e., territories free from the hegemony of the CFR.

The question is finding synergies stemming from cooperation to produce, distribute and consume based on agroecological experiences and the organized incorporation of new ones. The main objective of the ALOFOODs is to expand and supply the local consumption of healthy food, cultivated sustainably within the territory, applying fair work pay, and making consumption accessible in terms of price and physical location. These systems are on the opposite side of local or territorialized food systems formulated by conventional economists. The latter systems concentrate around one or more fresh or processed quality foods for which territories have a comparative advantage and which compete in national or international markets. This approach, which underlies the differentiated quality approach and protected geographical denominations, is instrumental to the CFR; it facilitates homogenizing local products, subordinates local production integration to vertical networks and long channels and does not guarantee an improvement to value added retention (Bowen & De Master, 2011; López-Moreno, 2014). From an environmental viewpoint, it does not bring about any notable improvement, because it does not contribute to reducing the metabolic or production profile, or the distribution or the reorientation of consumption (Edwards-Jones et al., 2008; Darnhofer, 2014). Conversely, the ALOFOODs are shaped to address local demand to the fullest extent possible, generating food sovereignty and placing the process at the heart of a self-centered local development strategy, generating a greater amount of added value and employment.

In this sense, ALOFOODs follow a double strategy of downstream and upstream cooperation, bringing all the links of the food chain into play and functioning on the basis of the territory and productive capacity of local agroecosystems. The ALOFOODs thus emerge from the convergence of two ideas: on the one hand, the approach of Local Agrifood Systems that articulate the potential of social and ecological sustainability with the capacity of the territory (Marsden et al., 2000; Ventura et al., 2008; Goodman, 2009; Bowen, 2010; Bowen & De Master, 2011), and on the other hand, the articulation of the different agents involved in the local food chain in a common project based on their cooperation and own territory (Marsden & Sonnino, 2008; Darnhofer, 2015; Bui et al., 2016).

From an upstream perspective, ALOFOODs consist of finding links between productions to close the nutrient cycles and reduce direct energy consumption. It is within the reach of producers themselves to create production networks as well as seed and input exchanges. In the agroecological experiences, many of these practices are already taking place. Moreover, we already have concrete examples of the potential of peasant cooperation for scaling up analyzed by Mier et al. (2018): the peasant-to-peasant movement in Central America; the national peasant Agroecology movement in Cuba; the organic coffee boom in Chiapas and Puebla, Mexico; the spread of Zero Budget Natural Farming in Karnataka, India; and the agroecological farmer–consumer marketing network "*Rede Ecovida*" in Brazil, etc.

As we saw in Chapter 3, processing, packaging, transportation, and sales in shops, i.e., the distribution chain, is responsible for a substantial part of the primary energy costs of the global food chain, transport is thus a critical point (Infante et al., 2018a). Expanding and consolidating shorter and more sustainable distribution and marketing channels is the goal of the ALOFOODs downstream approach. The chain's territorial approach favors: the location of agroindustrial activities in areas close to farms; the grouping of producers to sell in common, organize production, regulate supply and ensure supply; and, of course, making it viable to establish minimum logistical infrastructures to make this possible. The approach also allows us to weave effective links between production and consumption, as well as alliances with other local non-food actors that make it possible to anchor agroecological innovations through stable transformations of local food regimes. Finally, the local orientation of the ALOFOODs helps changing the typical consumption patterns that support the current regime: rooting in tradition facilitates a transition toward a healthier diet with less processed food and less animal proteins; a diet based on a greater amount of fresh seasonal products consumption than on highly processed foods coming from far away and overly costly in energy.

These local foods are today more expensive than they should be, precisely because they lack regular organizational and logistical support that reduces the costs of structuring the supply. The elimination of long and expensive processes, characteristic of long chains, will definitely result in lower end prices. In this sense, collective canteens, whether in public or privately organized centers, is an interesting lever to start building such circuits. In fact, introducing organic food in public centers (hospitals, schools, institutes, universities, barracks, etc.) has major carry-over effects (Friedmann, 2007; Izumi et al., 2010). In addition to providing healthy and waste-free food to users of these services, it is a powerful tool to educate about food and disseminate the virtues of organic foods among patients and their relatives, school children, parents of students, etc. However, it can also be a precious tool to stimulate production and short channels if priority is given to the supply to small and medium organic producers located near the catering centers. Experiences in Brazil and Andalusia (Spain) illustrate this (see Chapter 7).

This territorial organization of the food chain obeys the same criteria as that of institutional design described in the previous chapter. As is widely known, the more agroecosystems resemble ecosystems in structure and functioning, the more they are sustainable. Biomimicry (Garrido Peña, 1996; Gliessman, 1998; Riechmann, 2006) is an organizational principle applicable to ALOFOODs design, in which maximum connectivity and linkage with the territory is sought as well as maximum autonomy with respect to markets or the State or global chains. The territorial link is fundamental, not only because it seeks the maximum coupling between food and food production at local levels, but also because the territory gives meaning, provides identity and cultural significance to the very act of feeding, facilitating an anchoring in agroecosystems. In this sense, a territory is understood as a specific context for local development initiatives, that is, as the realm in which specific innovations are concentrated, reproduced and interconnected through "networking" and "institutional" types of anchoring processes generating radical and stable reconfigurations in local food regimes (Elzen et al., 2012; Darnhofer, 2015).

In short, the ALOFOODs are arranged into rural districts supported by cooperation and not global market competition based on products of differentiated quality. They seek strategic complementarities of economies of scale and, above all, scope economies for cost reduction, horizontal integration and the relative de-commodification of goods and service exchanges in the chain. They are directed toward internal markets and not export, seeking autonomy or food sovereignty through biophysical and cultural connections with the territory. They become agents of collective action and reflect social self-organization processes, that is, processes of articulation between actors and territorial resources that are sometimes concealed or snatched by hegemonic actors (Petersen et al., 2013). Although not essential, the social mobilization brought about by the construction of the ALOFOODs must involve public administrations, especially local ones (municipalities), that until now have been little involved in food policies such as health, education, the environment or territorial planning.

To summarize, reconverting conventional practices into agroecological practices throughout the food chain must continue to be the key of the scaling up process, but the practices must be directed toward the scale-change of the number of experiences. The main objective should be that of interconnecting experiences, enhancing the possible biophysical synergies between farms, ordering the territory to make it possible to close the cycles, cooperating to jointly produce inputs, to exchange seeds and improving them in a participatory manner, etc. The underlying criterion should be that of maximizing productive autonomy and, consequently, reducing market dependence, de-mercantilizing the exchanges or reducing them to a local level. This implies redesigning production on a larger scale than the farm, at the scale of a territory or landscape, reinforcing agroecosystems' internal circuits and their interconnections so that they become low-entropy dissipative structures (Guzman Casado & González de Molina, 2017). That is not all. A scaling-up strategy should foster the connections between production and other links in the food chain, promoting joint marketing and direct contact with consumers, involving the setting up of the necessary logistics. Logistics along agroecological criteria are key to building local food systems and massifying agroecological experiences. Within the framework of many agroecological movements and their experiences, practices have emerged which undoubtedly constitute a very useful tool for the escalation process, as highlighted by Mier et al. (2018).

But the question goes beyond the sole innovation and building of alternative food systems. The scaling up will not be possible without the impetus of social movements. The main task is to strengthen the movements themselves so that they gain in breadth and social relevance. The politicization of food consumption has demonstrated its great capacity to mobilize growing sections of the population. This is the most effective way to obtain social majorities to regain food sovereignty, now in the hands of the CFR. The following chapter is dedicated to developing this proposal of *food populism*, broadening the meaning of food sovereignty. In any case, the politicization of consumption is indispensable to boost the regime's necessary transformation, especially in highest income countries. Finally, the scaling up requires Agroecology itself to advance as a science and focus on aspects left unaddressed until now. One of them is Agroecology of the landscape, that is, the design of sustainable agroecosystems at bigger scales than the farm; or methodologies for the economic and environmental assessment of logistics and alternative channels. Not to forget two fundamental issues: on the one hand, a deepening of the feminist perspective on Agroecology and on the other, the strengthening of the axiological, theoretical and epistemological principles of Agroecology to face the cooptation attempts by the CFR and the institutions that serve it.

6 The Agents of the Agroecological Transition

According to the statistics cited in Chapter 3, family farms (including peasants) constitute over 98% of all farms, and work on 53% of agricultural land, producing 53% of the world's food (Graeub et al., 2016, 1). They constitute the basis of the world's food and the maintenance of global agroecosystems. It would seem natural that any strategy to advance an agroecological transition, massifying agroecological experiences and building an alternative regime be based above all on peasants. However, the corporate food regime (CFR) not only harms peasants, it "affects marginalized races, nationalities, genders, and classes of people, [so that] natures must be restored with the consent, participation, and design of those so affected" (Garvey, 2016, cited in Cadieux et al., 2019). Women, new farmers and consumers suffer the impacts particularly harshly. Without them, it will be difficult to achieve change in large parts of the world, a world that is already more urban than rural.

Consequently, scaling up agroecological experiences can only be done by mobilizing a social majority, led by peasants – whether traditional or "new" – in a global struggle for food sovereignty. Merely adding up claims, which are already fragmented and even contradictory, will not be sufficient to cement such a heterogeneous social alliance. A holistic political proposal is necessary, capable of promoting changes in production as well as in distribution and consumption. We called this proposal *food populism*, in accordance with other approaches (Cadieux et al., 2019). This alliance transcends the countryside–city dichotomy, that capitalism has used and that has underpinned its progression in agriculture. Food populism is the only way cooperative and solidary exchange between the two extremes of the food chain, the basis of a sustainable food system, will be possible. Despite its transversal nature, the food populism proposal has a powerful anti-capitalist element therefore also a "class" and gender component. In the following sections, we describe the theoretical and political foundations of this great alliance for change.

6.1 "FOOD POPULISM": BUILDING SOCIAL MAJORITIES OF CHANGE

Populism has a terrible reputation in the political and academic world. Populist movements have challenged the institutional order proper to western democracies and have thus led a large part of the academic world, especially liberal currents, to reject them wholesale. For authors like Shils (1956), populism designates a multi-faceted phenomenon underpinning Bolshevism in Russia, Nazism in Germany, McCarthyism in the United States, etc. For its part, Marxist critique has considered that populism is an ideology that, with very few exceptions, obscures the class dimension of social phenomena by appealing to the "people", thus generating a form of mobilization that can hardly favor revolutionary change.

However, another populist tradition exists that has not constituted a negative phenomenon, and it is perhaps no coincidence that is rooted in the peasant past and in some recent Latin American experiences. The term "populism" did indeed emerge as a political current in the mid-nineteenth century in Czarist Russia. Russian populists believed that the moral agency represented by the commune and its potential institutional adaptation based on modern agrarian cooperatives could represent the lever allowing to leap to socialism without having to descend into capitalist hell. The fact of "going toward the people" explicitly acknowledged peasants as the main subjects of the revolution. Critical intellectuals had to merge with the people to develop with them, on an equal footing, the forms of solidary cooperation that would lead to forms of progress that incorporated justice and morality (see a summary in Sevilla Guzmán, 1990; González de Molina & Sevilla Guzmán, 1993a).

Years later, Alexander Chayanov would offer the fullest pro-peasant approach to populist tradition (Chayanov, 1925 [1966a, 1966b]; Ploeg, 2013), based on a fertile combination with the Marxist theoretical tradition. His writings gave rise to an original synthesis that we have called *Neopopulism* in other works. In keeping with classical Populism, Chayanov recognized that the peasantry presented an anti-capitalist and socialist potential that was rejected by traditional Marxism. With this he recognized a multiplicity of subjects acting toward social emancipation, a task previously reserved only to the proletariat. Any social group objectively "confronted" with the system could – based on its own social conditions – contribute to social change without subordinating itself to the leading role of a single class with a revolutionary capacity. Chayanov affirmed that peasant solidarity and logic should drive alternative forms of development, in which technology should adhere to local cultural frameworks (González de Molina & Sevilla Guzmán, 1993b).

Late nineteenth-century peasant mobilizations that took place in the USA around the People's Party are commonly referred to as an example of populist movement led by poor farmers and that promoted progressive and anti-elitist ideas. The term would not become fashionable until the 1960s and 1970s with the emergence of national liberation movements and some government experiences in Latin America and other peripheral countries. In this context, rising theoretical–political movements used populism to mobilize large swathes of the population targeting social change; this form of mobilization was led by a charismatic leader and was based more on the emotional components of the mobilization itself than on the rationale of its claims. The post-Marxist proposal of Ernesto Laclau (2005) arose in this context. He grounded populism into a class ideology with a clear democratic content. Our proposal of food populism is also based on his contributions.

Populism for Laclau (2005) should not be understood as an ideology but as a political language of communication, as a form of politics that can contain authoritarian proposals from both the right and left. "Populism does not define a movement's ideology, it defines a type of political construction based on society's division into two fields [...] Emancipatory politics must necessarily have a populist dimension, but it must also be defined by the contents of that policy, not simply because it is populist" (Laclau, 2009, 826). Its main mission is to mobilize the "people" oppressed by a privileged minority. The people are not thus considered to be a truly really existing social category, but a plural category seeking to reduce social complexity. This is especially relevant in postindustrial societies where social fragmentation and the diversity of interests, even when they are found, make it very difficult to make claims based on class antagonism.

As is generally known, traditional classes have become more segmented since the seventies (Beck, 1998), generating diverse antagonisms, that have even confronted each other. Such antagonisms have been reinforced by the rising social entropy brought about by the economic–financial crisis; this entropy can no longer be easily compensated by a growing metabolic or biophysical dissipation. In fact, the increase in physical entropy has turned in recent decades into a zero-sum game, in which what some gain others lose (see Chapters 2 and 3). In this fragmented social framework, building social majorities of change based on the sum of the demands of the parties is difficult. As Laclau upheld, "the total lack of coordination is not a political solution either, and the idea of the multitude is the idea of an unstructured whole in which different forms of antagonisms begin to proliferate ... the antagonisms are much more complex than what the Classical Marxist theory presupposed ... the important thing is to experiment with new forms of articulation" (Laclau, 2009, 820). Paradoxically, the social complexity and the different forms of domination in these types of societies create favorable conditions for a wide range of conflicts and protests to emerge that can be coordinated by attributing their ultimate responsibility to the institutions that make them possible, i.e., the elite that benefits from it.

Populism is therefore a political language capable of articulating diverse interests under a united mobilization against the system, unveiling the basic contradiction between the "people" (the social majorities) and a privileged minority. In line with Laclau (2005), the articulating role of fragmentation lies precisely in the building of global antagonism, capable of creating the subject of social change through mobilization. This unifying and emotional discourse can grow more easily in the political terrain of democracy and identities. It is in that latter arena that the discourse can succeed in generalizing protest and challenging the cultural and political hegemony exercised by the ruling class, that is, the elite. The reason is that populism relies on a powerful imaginary democratic–egalitarian construct, proper to postindustrial societies and on the articulation of a broad block, a widespread social majority, as the only path toward democratic change. That is, populist mobilization is the only way to face the diversity of demands, that join together to oppose the dominant classes, appealing to democratic radicalism and the recovery of sovereignty in the hands of the people. Therefore, we believe that the concept of populism is of an instrumental nature, and naturally, in accordance with Laclau, a positive one. Populism can be understood as a grammar that produces the people, a historical subject of change (Retamozo, 2017, 170).

When it comes, as in our case, to defending the rights and interests of peasants, farmers, and the vast majority of consumers generally, i.e., of the "people" against an "extractive and exploitative" elite, populism imposes itself over any other more fragmentary, therefore more complex, political language of communication. The sphere of conflict generated by the CFR is a political playing field that is particularly adapted to this type of language due to its socially transversal and geographically global nature. The interests and demands expressed along the food chain are varied and even divergent: producers and distributors, some territories versus others, between the countryside and the city, between producers and consumers. The conflict is embodied in disputes around end prices of food or cultural disparities between rural and urban areas. The transformation of the CFR requires producers and consumers to form an alliance, making it possible to mobilize the majority of the people against the elite governing it, pressuring public institutions (at subnational, national, and international levels) to change the institutional framework. Going further, this alliance is essential to build an alternative regime based on direct contact and trust between producers and consumers. Moreover, the regime can only be defeated on a global scale and this requires alliances between the north and the south, where apparently highly disparate social groups must join forces against the common adversary.

Populist language is undoubtedly the vehicle of this essential union. Language allows emphasizing points of unity raising them above points of contention along the food chain. The proposal is that of populist mobilization around the supreme values of our species (health, equity, respect for the environment, etc.) that also include emotional elements as well as demands for democratization – expressed by vindicating food sovereignty and all that it entails. In that sense, almost 99% of the population is potentially against a CFR that is directly responsible for hunger, undernutrition, malnutrition, rural poverty, structural unemployment, and major damage to health and the environment. As we will see below, naturally, we would reject any resulting antidemocratic social and political movements, but not anti-liberal ones. In that sense, democracy is not associated with liberalism but with republican culture. Consequently, populism can lead to anti-liberal movements, while being radically democratic.

To summarize, our food populism proposal is based on two converging intellectual traditions and practices: Russian populist and neopopulist tradition, and the more recent post-Marxist theoretical–political and academic tradition. Our reinterpretation some time ago of the Russian populist tradition included the environmental component and we formalized it in our proposal of "ecological neopopulism" (González de Molina & Sevilla Guzmán, 1993b). The latter proposal is based on the material or physical sustainability of peasant production forms, that are isomorphic with respect to future sustainable agriculture. However, the proposal is also based on the suitability of the cooperative and intergenerational basis of peasant social relations allowing to regulate the trade-off between social entropy and physical entropy. In this sense, food populism can be considered from an evolutionary perspective as a tool to promote altruistic behaviors of multilevel selection in which the individual is sacrificed for an abstract ideal (demos) and individual welfare is replaced by forms of genetic, reciprocal, or spatial altruism (Nowak, 2006). As such, the tool is neutral, it has no specific underlying political purpose, but is an ideal instrument during historical phases of austerity. It can be used to make periods of coercive austerity bearable in contexts of increasing inequality, e.g., during economic recessions and rising conservative or authoritarian populism, but it can also be used to stimulate austerity in times of decline; for example, during the agroecological transition toward a sustainable regime. Our food populism proposal thus shares the same ecological and evolutionary grounding as Political Ecology generally.

6.2 PEASANTS: CENTRAL ACTORS IN THE AGROECOLOGICAL TRANSITION

The peasantry must, for obvious reasons, be at the heart of populist mobilization as we have defined it, but beyond their weight in agriculture and food globally, peasants possess traits, knowledge based on millenary practice, and generally manage agroecosystems, making them ideal key players of the agroecological transition. Peasants, their families, and their communities have been the most affected by the CFR. Nevertheless, peasants present a broad range of situations from an agroecological perspective: from those who are more or less involved in the industrial management of their farms, often in a forced way, to peasants who are leading the struggle for sustainable agriculture. Therefore, Political Agroecology needs a theory of peasantry that reflects this diversity. That is, a theory that explains not only the peasantry's social nature and historical evolution, but also its subordination to entrepreneurial logic, and the loss of peasant identity. Such a concept is essential to design agroecological strategies that reverse the process. We will start by characterizing the peasantry from an agroecological viewpoint and finish with a theory on the degradation of the peasant condition.

Bountiful definitions have been given of the peasantry (see a recent state-of-the-art in Bernstein et al., 2018). The problem may come from the so-called "Peasant Studies" school, an area of sociology, and its quest for a conceptual category embracing the huge variety of situations found in different countries after the Second World War. This issue has led to endless and confusing controversies about whether the peasantry constituted a class in its own right or only a part of a larger society structured into classes; whether, as a class or group, it belonged to an outdated production regime (such as feudalism) or whether its survival was worth considering under capitalism; whether it constituted a mode of production or was only a part of a larger society; whether the most suitable term was peasant, family farmer, simple producer of goods, etc., and what the substantive differences between such denominations could be. The whole debate arose from the increasingly blatant reality that peasants had not disappeared, despite the prophecies of classic agrarian social thought and even that of the most liberal sectors. It was necessary to define a category that could explain both its survival and its transformations. Shanin (1979, and especially 1990) pointed out how absurd it was to attempt to accurately define a social group that had always existed and continues to exist. The implications of the discussion are not purely theoretical. The debate continues and has even acquired a new dimension with new environmental variables and its major role in the fight for an alternative regime (Calva, 1988; Toledo, 1994, Kearney, 1996; González de Molina, 2001; González de Molina & Sevilla Guzmán, 2001; González de Molina & Toledo, 2014; a revision in Bernstein et al., 2018; Ploeg, 2018b).

However, the emergence of the food sovereignty paradigm and the peasantry's leading role has not led to any agreement on a clear concept. There is still considerable confusion around the definition of categories of small farm owners. Some still refer to peasants to allude to family farmers, others instead call them petty commodity producers, most identify the peasantry only with family farms and eventually use this concept to get around controversies; incidentally, this concept fails to reflect many changes and different situations underlying such a generic denomination. In short, no theory accounts for the peasantry's major defining transformations and their explanation.

The definition we put forward next is a contribution to solve this controversy. Our aim is only to approach the notion of peasantry from an agroecological perspective. To do this, taking ecological aspects into account is not enough, it is also necessary to closely relate these aspects to socioeconomic and cultural elements. The best way to advance in this task is to adopt, in accordance with what we saw in Chapter 1, a metabolic perspective. This theory uses many ideas proper to classic thought, although seeking their socioecological rationale or advancing explanations based on that rationale. In line with Shanin, as mentioned above, a peasant can be defined as a person who practices peasant agriculture, i.e., agriculture performed following specific economic–ecological rationality and distinct from capitalist agriculture. Their features are not static, they evolve according to time and space dimensions, conforming to the features of the agroecosystems and the conditions of the societies in which they live. Hence the great diversity of peasant agriculture found across time and space. Thus, it seems logical to define peasantry as a social category that is essentially historical, whose features have maintain a certain unity though they have been undoubtedly transformed over time. Attending to its socioecological nature, we may consider peasantry as: "owning a fragment of nature that is directly appropriated at a small scale by means of their own manual labor, having solar irradiation as their fundamental source of energy, and their own knowledge and beliefs as their intellectual means. Such appropriation constitutes their main occupation, from which they consume firsthand, in whole or in part the obtained fruits, and from which they either directly or indirectly, through their exchange, satisfy the needs of their families" (Toledo, 1990). The latter perspective allows identifying peasantry as a social category associated to one of the forms of articulation of social metabolism: peasants are those who practice peasant agriculture, that is, they carry out organic metabolic management of their agroecosystems. In other words, peasantry is the social group around which agrarian societies organized – and continue to do so in many regions of the world – the agrarian activities in societies based on solar energy. That implies establishing a rather strong identification between *organic metabolic regime* and peasantry (González de Molina & Sevilla Guzmán, 2001; González de Molina & Toledo, 2014; Petersen, 2018).

Indeed, most of the defining features argued by the "tradition of peasant studies" were either "functional" or highly adapted to types of organically based economies that were by their very nature steady-state economies (Daly, 1973; Tyrtania, 2009). Most of the defining features (Calva, 1988; Sevilla Guzmán & González de Molina, 2005) were functional or were well-adapted to organic-based economies, which could only function with producers that could establish an identity between agrarian production with family economy, and which could mobilize all workers that could be available for agricultural labor, developing successional and marriage strategies that aggregated the factors of production as much as it was possible, thus ensuring the utility of appropriation for the survival of future generations. The only way organic economies could function was through the existence of a farmer mutual support network, mediated through kinship, neighborhood, or friendship, in such a way that families were guarded against adversities. Their functioning depended on the generation of common culture and ethics, on an identity that gathered and codified knowledge about the environment, the crops, the practices of animal management, the successful or unsuccessful practices of facing everyday challenges, etc.; that is, economic functioning depended on all that was indispensable for sustaining successful farming through the years. They could only function through a multiple use of the territory, taking advantage of the necessary spatial heterogeneity imposed by the complementarity and integration of agricultural, livestock, and forestry land uses, the only way to make agroecosystems functioning of the agroecosystems. Multiple land use was also a strategy of diversification of the inherent risks of climatic or economic fluctuation, such that its maintenance in good conditions – for example, the respect for natural cycles and the systems of soil fertility restoration – became a *sine qua non* condition for peasants' subsistence and the future survival of their children. It is for these reasons that the existence of an ecological rationality has been proposed (Toledo, 1990; Toledo & Barrera-Bassols, 2008).

By this we do not mean that these types of societies have not known environmental failure or ecological crises. They have, and they sometimes caused the socio-ecological collapse of the societies involved (Tainter, 1988, 2007; Diamond, 2004). What we mean is that peasants depended essentially on exploiting natural resources, their livelihood was based more on nature's products than market products (external inputs). They were, therefore, the first people to be interested in using agroecosystem formulas guaranteeing an uninterrupted flow of goods, materials and energy in a natural way; whether they achieved this or not depends on the analysis of each particular society. Therefore, no immanent idea of environmental good exists in the peasantry. Deterioration translated more quickly into moral or social penalties and even into additional market costs than today.

With this multiple use strategy in mind, peasants achieved their social reproduction through manipulation of the geographical, ecological, biological, and genetic components (genes, species, soils, topography, climate, water, and space), and the ecological processes (biological succession, life and astrological cycles, and material and energetic flows). The same diversified arrangement applied to each of the productive systems, e.g., terrestrial or aquatic multiple-cropping systems instead of monoculture, monospecific herds, forests, or fisheries. In sum, peasant units tended to engage in unspecialized production based on the diversity of resources and practices. This mode of subsistence and reproduction promoted using all surrounding landscapes to the maximum degree, the cycling of materials, energy, and wastes, the diversification of obtained products, and above all, the integration of different practices: agriculture, forest extraction, agroforestry, fishing, hunting, small-scale animal husbandry, and crafts. This is a kind of natural pluriactivity (Toledo, 1990), far from the work that family farmers are forced to do in other non-agricultural activities, due to the lack of sufficient income under the CFR.

All peasant producers need intellectual means for appropriating nature. In the context of a peasant economy, such knowledge of nature became a decisive component in the design and implementation of autonomous reproduction strategies. Such knowledge was orally transmitted across generations, and through it, peasants refined their relations with the environment. Because this body of information follows a logic differing from that of current science, it has been called *saberes*, a Spanish and Portuguese term referring to a pool of particular (local) knowledge, wisdom, and know-how (Toledo & Barrera-Bassols 2008). Peasant societies own a repertoire of ecological knowledge that was generally local, collective, diachronic, and holistic. In fact, peasant groups have, over time, developed management strategies and generated cognitive systems about their surrounding natural resources, which are transmitted across generations. But peasants not only accumulate practical knowledge. Through experimentation, combining management with intellectual work, they also develop a remarkable capacity for productive innovation.

On the other hand, the stability of the organic metabolism is based on the continuity of the flows of energy, materials, and information. A number of social institutions tried to ensure the maintenance of the group, protecting society from environmental or economic perturbations, its resilience and stability through time also depending on the efficiency of such institutions. The role of domestic units is well known in the context of human reproduction and peasant economies, and in developing strategies that eventually affect the size of the population and the capacity for generating productive work. For this reason, we will not deepen the subject further as it has been sufficiently covered by historical and anthropological works (Goody, 1986; Bourdieu, 1991, 2004).

However, the peasant group could only handle a small portion of suitable land, i.e., the land it exploited itself. Joint management and control was essential for the peasant community – and also for the survival of the domestic economy itself. Communities are the smallest units of a territory's population and in them peasant domestic groups predominate, specialized in agrarian activity, toward which all other crafts or professional activities are oriented – as auxiliary, complementary, or dependent activities. Given the localized and usually relatively closed dimension of energy and material flows, the scarcity of external exchanges, and land population and distribution patterns, most social life took place in hamlets, villages, or small population hubs. The information essential for the functioning of social metabolism flowed in these latter localities. The peasant community is, therefore, the minimal unit of production organization driven by solar energy, because peasant domestic groups are not capable of individually owning each and every condition for production (pastures, forests, etc.). These characteristics are particularly significant regarding the new territorial and social configuration of a sustainable food system. Cooperation, mutual aid, integration of the different uses of the territory, etc., are essential to close cycles and for productive autonomy to be possible.

From an institutional perspective, the peasant community has ample competencies upon all factors of agrarian production and of the whole of the appropriation process. Their political institutions – whichever these were – were responsible for instance for establishing norms impeding overexploitation of forests or the land, or overgrazing during the appropriation of firewood or manure; they were also responsible for regulating changes in land use within the territory, fomenting or not the necessary equilibrium needed between the different exploitations within it; they were also in charge of guarding the personal conditions of production, through the execution of actions in the context of public health, charity, instruction, defense against external threats, or provision of material aid in moments of crisis. Overall, the elaboration of collective norms in societies with organic metabolism aimed at avoiding both the use, and the excessive consumption of common resources (Warde, 2009, 76). All these collective norms regulated, sometimes rigorously, farmers' functioning and practices. Nothing is more distant from the old ideas of the tragedy of the commons put forward by Hardin (1968), identifying the open fields and the communal system with unrestricted access to them. Both access and use were subject to strong regulations seeking cooperation and preventing free-rider behaviors, i.e., that no neighbors take advantage of common resources in such a way that these could be overexploited. In other words, these social institutions attempt to constrain social and ecological entropy – given that the transferences from and to other territories used to be limited – and to distribute it equally among all. In fact, as soon as the new capitalistic social network is established these social institutions become progressively superfluous or are even abolished by the new legal regime shaped by capitalist rationality. As stated by Paul Warde, "the abolition of systems of collective constraint through the enclosure movement, that waxed most prominently and early in England but that would embrace most of the Western Europe in the nineteenth century, remove the issue of the consequences of neighbourly action from the agronomist's purview. The system of agricultural action became the farm unit, the environment denoted as 'natural' or 'market' forces" (Warde, 2009, 76). Laws and other norms, either positive or consuetudinary, had a first-order stabilizing role. In Chapter 4 we saw that from the point of view of Political Agroecology, the communal property of natural resources in its broadest sense is extremely useful for designing and establishing an alternative food regime.

All the features described above characterize the peasantry in a pre-modernization version of agriculture (employment of family labor and renewable energy, multiple use of the territory, local resources, autonomy with respect to markets, cooperation and mutual aid, commonality, traditional knowledge, innovation and adapted technologies, closeness food production and consumption proximity, etc.). They consist of ways of managing agroecosystems, types of social relations, and institutional arrangements included within Agroecology proposals to shape a sustainable food system. As we have seen in the previous chapter, an understanding of peasants as subjects or protagonists of the agroecological transition and the alternative regime is based on these traits.

Nevertheless, globally, in the real rural world, there is a huge diversity of situations that reflect to a greater or lesser degree the peasantry's defining features. The thesis we defend maintains that this diversity results from peasant traits degrading until practically disappearing or turning into other related social categories, proper to industrialized agriculture. This does not mean that these categories have definitively lost their conditions of peasantry and that this degradation is not reversible. The agroecological transition offers precisely the opportunity to reverse this process. Classical theories analyzed the development of capitalism in agriculture based on the competition between small and large farms. The process ended with the disappearance of small farms and the definitive predominance of large farms and salaried work (see Chapter 5); but these theories only partially capture what happened and perhaps only the least relevant part. We do not believe that the main reason is farm size and competition between farms in the agricultural sector. We believe it is better to understand the *degradation* or *deactivation* of peasant traits owing to Capitalism penetrating the gradual commodification of peasant production and reproduction strategies, fostering a distinct resource management, i.e., an industrial one, that peasants progressively understood they had to use if they were to survive.

6.3   PEASANT CONDITIONS UNDER CAPITALISM AND AGRICULTURAL INDUSTRIALIZATION

Indeed, in hindsight, the agricultural sector's evolution does not fit the forecasts of the classics. The announced differentiation and proletarianization has not led large farms operated by salaried workers to become the dominant form of organization of agrarian industrial production. Family farming, as we have seen at the beginning of this chapter, is the prevailing form of organization of agrarian production in the world. Although certain features proper to family farmers or small producers of goods remind us of the peasantry (the ownership of a small farm, the family's contribution to labor, etc.), many of these features have been blurred. This process, however, is dynamic, and has even brought about opposite trends, as we will see later, leading to the emergence of farmers who seek to reduce their market dependency and strengthen their autonomy. They have been denominated "new farmers".

The origin of the loss of peasant traits lies in private property and the market penetrating into peasant economies, causing a sudden leap into commodification and subordination to the capitalist market. The process began in Europe with liberal reforms. They destroyed the traditional system of open fields and communal use, based on integrated agrosilvopastoral use, by means of enclosure laws, private appropriation of property and communal rights and the consideration of land as a commodity. This led to a powerful process of peasant dispossession and the destruction of the peasantry's material foundations, that used to make it possible to close biogeochemical cycles on a local scale (that is, reproducing fund elements of agroecosystems without resorting to external inputs; Guzmán Casado & González de Molina, 2017). These new circumstances led the peasantry to redefine their reproductive strategies: many goods required for subsistence became goods that could only be acquired with money and via markets. Peasants were thus impelled to specialize their production and to seek higher yields from their small plots. It became increasingly difficult for peasant farms, owing to their small size, to apply traditional integrated livestock agricultural systems (use or ownership of pastures had been privatized forcing them to acquire them through the market or, more frequently, to dispense with livestock). In this way, they were forced to increase their participation in economic market flows while reducing flows with nature, converting products that were previously considered as values of usage into objects of change.

The whole commodification process underwent a spiral growth, accompanied by lower agricultural prices for peasants. The progressive integration of international agricultural markets and the added value differential between agrarian and industrial production pushed the monetary remuneration of harvests downwards, and in fact continues to do so today. The answer was, as we saw in Chapter 2, to intensify the use of capital and productive specialization to compensate the loss of income with bigger production volumes. As cycles could no longer be completed locally, due to loss of rights and communal goods, peasants were caught in a vicious circle of growing market dependency: they had to increasingly resort to purchasing ever more expensive external inputs to increase production, receiving, in return, prices that kept dropping. They thus found themselves involved in new markets, in this case in input markets (fertilizers first, implements next, ultimately transferring a large part of their income out of the farm through the purchase of machines, fertilizers, improved seeds and phytosanitary products or in interests on loans to face growing investments) and eventually depended on these markets to obtain a sufficient harvest. This phenomenon applied to the strictly productive field, but it affected household economies in a similar way: the disappearance or privatization of communal property and rights channeled a major share of peasant subsistence (food, fuel for heating and cooking, clothing and accommodation) into the market, subjecting peasants to greater work efforts to obtain the money necessary to have access to those goods. A situation whereby peasant household reproduction depended on agroecosystems turned into a situation where reproduction became dependent on markets (Ploeg, 1993). The reproduction of family farmers thus depended more heavily on the existence and affordability of flows of nutrients (fertilizers), defense against pests and diseases (pesticides), and fuels (diesel or electricity) for machines and tractors, than on the environmental quality of their plots and that of the surrounding environment.

The process of commodification of agroecosystem management eventually subordinated peasantry to capitalism, converting agroecosystems into providers of cheap food. The system, through market regulations and public policies, outsourced the real cost of food production while it failed to compensate total reproduction costs related to peasant households and their farms. The key to the peasantry's survival under Capitalism has perhaps been the flexibility of peasant holdings, that are capable of assuming low prices for its products, favored by the system's traditional difficulty in fully industrializing types of production that are heavily dependent on natural cycles. The "formal subsumption" of the peasantry into capitalism (González de Molina & Sevilla Guzmán, 1993a) has been a cheap way to secure inexpensive food for all the other economic sectors; that is, it has constituted a steady source of capital accumulation. The peasantry's land exploitation by markets, to a much greater extent than the exploitation of landless peasants' wage labor, has been the basis of cheap food and work, two of the "four cheaps" referred to by Moore (2015), and the basis of the expanded reproduction of capital.

This commodification process has been slow, of unequal duration according to the rhythms of economic growth itself and materialized via two parallel processes: on the one hand, through the agroindustrialization of peasant agroecosystem management, and on the other, through the gradual erosion of the peasantry's cultural, identity, and consumption patterns. The process could be studied as the degradation or deterioration of the peasant condition. Víctor Toledo (1995) attempted to illustrate precisely this point by studying the degree of peasantry of family producers in Mexico.

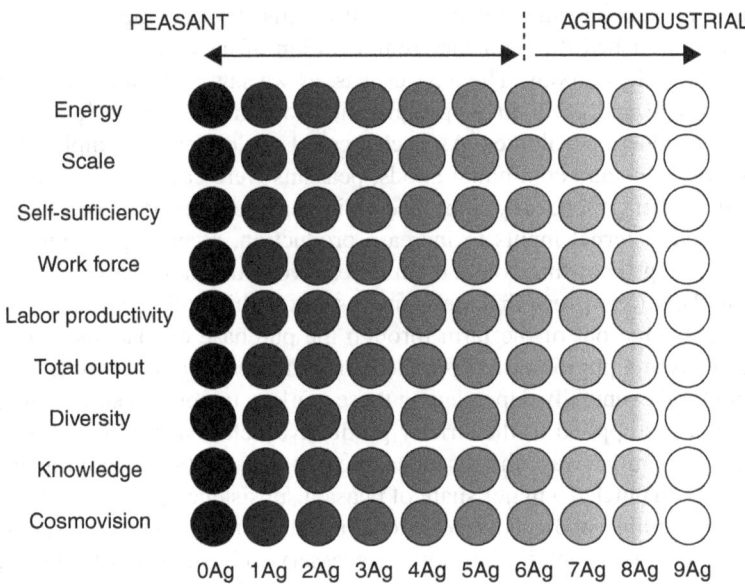

FIGURE 6.1   Degrees of peasantness according to Víctor Toledo (1995). *Source*: Altieri & Toledo (2011, 594).

Although the author uses the concept of "degrees of peasantness" (Figure 6.1) to measure the survival of peasant traits among small Mexican producers, his reasoning is perfectly applicable to the transition between the organic metabolic regime to the industrial one, that is, to the process of industrialization of agriculture and subordinate integration in the dominant regime: "Between the two archetypes defined above [that of peasant production and that of agroindustrial production] there is a range of intermediate situations resulting from different combinations of peasants' typical and agroindustrial traits. These combinations depend, in turn, on the 'moment' in the 'modernizing' process that tends to transform the peasant model into an agro-industrial model." In short, the degree of peasantness would be inversely proportional to the achieved degree of commodification.

In some cases, peasants have been transformed into "proletarians" or salaried workers in industry or services, in line with classical theory, through the constant transfer of labor from the countryside to cities; but others have remained, as shown by the still decisive percentage of peasantry remaining, above all, in the periphery of the developed or rich world. Many remaining family farmers in Western agriculture can be classified as direct descendants of peasants who lost, one generation after the next, many defining features. Some of them struggle to remain as autonomous and independent from the market as possible and must be considered as "new peasants". Building alternative food systems based on agroecological production, distribution and consumption criteria is the only way to reverse the degradation process. Likewise, only by reversing that process will it be possible to provide a solid basis for sustainable food system configurations. As claimed by Jan Douwe van der

Ploeg (in Bernstein et al., 2018a, 694), peasantries undergo a constant process of depeasantization and repeasantization. This implies definitively banishing the idea of an inevitable course toward disappearance, not only because of the CFR's objective interests, which use peasants as a way to continue appropriating cheap food, but also because of their own resistance to the mercantile and capitalist logic: in that sense and citing Ploeg once again, "peasant agriculture features as a promise for the future (instead of as mainly or just a remnant from the past)" (Ploeg in Bernstein et al., 2018a, 695).

6.4 THE "NEW PEASANTS"

We mentioned earlier the ongoing tendency to reduce the prices received by farmers practically all over the world. We saw that this owed both to the quasi-monopolistic pressure of the CFR, including the economic policies of governments bent on providing cheap food. Among the multiple responses, it is worth noting "repeasantization" processes both in the industrialized center and in the peripheries. According to Ploeg (2008, 7), we consider this process as "the fight for autonomy and survival in a context of deprivation and dependency [...] Repeasantization implies a double movement. It entails a quantitative increase in numbers. Through an inflow from outside and/or through a reconversion of, for instance, entrepreneurial farmers into peasants, the ranks of the latter are enlarged. In addition, it entails a qualitative shift: autonomy is increased, while the logic that governs the organization and development of productive activities is further distanced from markets". New farmers gain autonomy from markets by reducing costs, reducing the use of external inputs, and/or marketing their products through alternative channels. Their ways of managing farms and the household economy thus recover the ways of peasant management, intensifying work flows and knowledge as well as stimulating the internal circuits of agroecosystems, reducing the use of energy or nutrients coming from the outside. These sorts of strategies can be seen in many organic agricultural processes in Europe, Japan, or North America; they can also be found in many neorural movements, in peasant and traditional people movements linked to *Articulação Nacional de Agroecologia* (ANA) in Brazil (Petersen et al., 2013) or in the creation of new small farms in Pakistan, Bangladesh and India (Ploeg, 2008, 9). Finally, in line with the reversible nature of the peasant condition loss, repeasantization is also the answer given by many farmers to the pressure exerted in their countries by the CFR.

Nevertheless, repeasantization not only consists of changing productive strategy and bringing farmers closer to the peasantry. We again agree with Ploeg, "repeasantization implies a revival of values – autonomy, self-provisioning, local links and citizenship – which often have a strong capacity to mobilize large sections of today's societies" (Ploeg in Bernstein et al., 2018, 697). This is a fundamental aspect of the food populism proposal, as we will see later. Only a broad repeasantization process in both the productive and cultural sense could cope with the CFR's current crisis and build truly sustainable food systems (Ploeg, 2008, 11).

However, repeasantization, in line with its nature of process as described, means a threat to the CFR. The institutional framework that governs its functioning and that guarantees its reproduction is tasked with hindering the progress of alternative agriculture by making it conventional. We can define *conventionalization* as the process by which non-business production (peasant, family production, or non-commercial organic production, etc.) ends up subjected to the market, that is, dependent on the market to reproduce itself. In this way, Capitalism appropriates the alternative (non-capitalist) aspects of organic or peasant agriculture and uses them to generate monetary benefits (accumulation). The process, which may be involuntary, involves all food practices, because the rules of the game in which they operate push toward market dependence.

As we mentioned in Chapter 5, the case of organic production (whether certified or not, whether in the hands of family farmers or not) is a good example of this process. It entails the proliferation of a production model that repeats the characteristics of conventional agriculture and food, reproducing the same history and sharing the same social, technical, and economic characteristics (Allen & Kovac, 2000; Rigby & Bown, 2003, Raynolds, 2004; Reed, 2009; see a review in Darnhofer et al., 2010; Petersen, 2017). The lower perceived prices, resulting from CFR pressure, also affects organic producers, and pushes them toward externalizing territorial costs (Guzmán Casado & González de Molina, 2009) further (fewer rotations, fewer crops, high-response seeds, more phytosanitary treatments, etc.) and, therefore, to depend more on external inputs. Organic producers, therefore, have a clear economic motivation to find shortcuts to economic viability, at the expense of sustainability. This tendency is favored by regulations (organic production regulations, for example) that allow these types of external solutions (for example, by the recurring penalizing of the self-production of seeds, plants or phytosanitary products). Therefore, if the institutional framework does not change, organic agriculture, tends to reproduce the same model as conventional agriculture, in addition to organic production segments that are openly part of the agribusiness. This also means that the degradation of the peasant condition cannot be reversed, nor will the process of repeasantization be successful.

Distribution is following a similar course. Organic production mainly circulates through the same commercial channels as conventional foods. They are dominated by long distribution channels that heavily consume energy and materials, that can cancel out organic production's positive environmental impacts or notably attenuate them. Organic producers are often forced to sell their products through large food companies that develop their own organic brands for off-farm work (i.e., processing, distribution, and sale). We must add that the imbalance between increasing demand and insufficient and poorly organized supply, as in the case of Europe today (EU-DG AGRI, 2010, 42), favors the entry of large distribution operators reproducing the same conventional model in which farmers retain a ridiculously small percentage of the final price. Organic production thus ceases to represent a form of resistance to the industrial model of food distribution. Consumption is behaving in the same way, because dietary patterns do not change only through organic food intake. In fact, green markets ensure that organic foods can almost completely replace conventional foods, but the relative prices of each fail to stimulate a diet change. All these processes, operating within the so-called conventionalization pattern, show how important the institutional framework is and the need to create new institutions to ensure that agroecological experiences follow the course of an alternative food system (ALOFOODs, for instance). Without institutional change, i.e., without building a new institutional set up, it will not be possible to scale up Agroecology or the sustainable reproduction of the new peasants.

6.5   AGROECOLOGY AND FEMINISM: THE CENTRAL ROLE OF WOMEN

As previously referred to, the gender perspective is essential to transform the food regime according to an agroecological approach. The sexual division of labor, characteristic of patriarchy, is one of the most primitive forms of inequality. In industrial metabolism societies, sexual division was reformulated based on the radical split between productive and reproductive work. The first is a source of value, recognized by the market and concentrated around extractive activities, while reproductive work is invisible and devalued because it is outside the salary–mercantile system. Women have had to carry the full weight of concealed and devaluated reproductive work, leading to a mechanism of domination that has still not been broken, not even with women's incorporation into the labor market throughout the twentieth century (Federici, 2018). The primitive accumulation of capital in industrial metabolism would have been impossible without slavery and women's hidden reproductive and care work. There are invisible variables within variable capital in the organic composition of capital: women's reproductive work that goes far beyond mere physical reproduction and extends to domestic tasks, to community care, functions of mutual support and cooperative coordination of non-monetarized social work. If we incorporated these hidden tasks into the variable capital, the capital equation would break down. In the same way, if we add metabolic costs to constant capital, the fantasy of stable organic composition of capital becomes impossible. Women have been confined to the field of reproduction and excluded from production. All this justifies the necessary coupling between Ecology and Feminism, a union that has produced political proposals that have been adopted by Agroecology as its own (Puleo, 2011).

Capitalist industrial metabolism societies have required two safety valves for their increasing levels of social entropy: an external valve, which compensates with increasing physical entropy, as indicated in the first chapter; and an internal valve, which tries to displace social entropy (inequality) toward the periphery (poor and dependent countries, slavery, peasants) and toward women, leaving jobs associated with social reproduction unpaid. Placing these jobs outside the labor market has helped to develop industrial metabolism and capitalism. The blockage of these two external and internal escape valves is closely linked to the current metabolic crisis. The multifactorial ecological crisis, financial globalization, industrial relocation, the emancipation of women and their incorporation into the labor market represent entropy-flow short circuits that are leading industrial metabolism toward a risk of collapse.

Ecofeminism has highlighted the existence of a close link between the unsustainable management of ecosystems and their resources and the predominance of patriarchal relations. In fact, one could say that the non-recognition of women's reproductive activities is clearly connected to the unsustainable management of natural resources. Anthropocentrism and androcentrism are two sides of the same coin from an ecofeminist perspective.

This crucial gender-based inequality indeed existed in pre-industrial societies, but it significantly intensified under Capitalism. We have already seen how Capitalism grounded its accumulation process in the cheapening of raw materials, food, paid work and the non-payment of a substantial part of the social work necessary for subsistence, the work involved in social reproduction. The essential alliance of Agroecology and Feminism derives precisely from how Capitalism rendered invisible the whole range of tasks and processes involved in reproducing the two elements involved in production: human labor and natural resources, that is, the fund elements. Capitalism outsourced these costs, exploiting human work and nature at the same time. Adopting a feminist perspective in Agroecology is not only necessary to eliminate gender inequalities, but also precisely to ensure that the fund elements of both a social and ecological nature are reproduced. Feminism and Agroecology must be inextricably linked to bring about the real costs (of reproduction) of productive activities, both family costs (care, etc.) and ecological costs (fund elements). That is why Agroecology cannot exist without feminism, as feminist agroecologists claim. In addition, it is essential that feminist and agroecological movements form strategic alliances to make Agroecology's upscaling possible and to build sustainable food systems, the latter adopting a gender perspective and the former an ecological perspective.

This alliance is necessary because the gender approach is of utmost importance to the agroecological movement: indeed, Agroecology does not propose a technological or legal change in the food production model, but a political change to the social reproduction model. Women have a broad historical experience and a wide range of community and cooperative social skills that can represent a collective intelligence bank to use during the agroecological transition. The gender perspective is extremely useful when managing agroecosystems; women's centuries-long practice of both social and ecological reproduction is not in vain. It is not that women possess an instinct that is favorable to conservation, but their dedication to these tasks for centuries makes them more sensitive to issues relating to social and ecological reproduction (Agarwal, 2010). In other words, women are carriers of a logic that is nondestructive of nature due to their historical exclusion from productive tasks.

Moreover, women are chiefly responsible for acquiring and preparing food; and they are often responsible for producing it directly even in agriculture, when caring for vegetable gardens, farm or backyard animals; and they usually also look after their conservation or transformation. Women also tend to take care of children's health and education, an essential channel to foster different consumption habits. Many women also work in the conservation and exchange of seeds or in knowledge transmission regarding the production and use of medicinal plants. Many of these activities, usually considered secondary activities that hardly generate any monetary income, are nonetheless fundamental insofar as they are part of reproductive activities of a socioecological nature.

The feminist movement, and especially the ecofeminist movement, is an objective ally of the agroecological movement, as it sheds light on the impacts that agroecosystem degradation produces on women. Rural poverty, hunger, or malnutrition are often the lot of women, especially in poor or peripheral countries. Gender discrimination makes women even more vulnerable to food and nutrition insecurity. Meanwhile, ecofeminist movements have raised claims about health issues deriving from the deterioration of agroecosystems as well as food contamination and its links with women's sexual and reproductive rights.

The alliance is also essential for an agroecological proposal to succeed, given that women's cause is a powerful instrument of social mobilization that potentially affects a majority of the population. In fact, the gender approach has gradually acquired decisive importance in the agroecological movement in such a way that it is not possible today to conceive an agroecological movement without a gender approach, in the same way that decisive issues such as health and nutrition or reproductive health itself cannot be properly addressed without adopting an agroecological perspective.

From an agroecological point of view, the most efficient firm model is the family farm for a number of reasons detailed in Chapter 4, including the propensity to establish mutual exchange relationships, the productive and reproductive work convention, identification with the territory or *locusphilia* (Garrido, 2014), and intergenerational stability among others. But to prevent families from becoming subcontractors of sexual division of labor exploitation, which unequally allocates roles, rights, and resources, families must be diverse, equal, and democratic households. The agroecological gender perspective and ecofeminism are essential in this sense. Otherwise, family businesses will only be what the large majority of so-called "family economy" firms are: low-cost subcontractors acting for big agricultural companies (Reher & Camps, 1991).

6.6 POLITICIZING FOOD

In the nineteenth century, the Russian appeal to the people addressed its vast majority, the peasants, in opposition to the noble elite. Today that appeal includes, as we have seen, the peasants, new peasants, women, and consumers generally. Building an alternative regime is not solely the responsibility of producers or distributors. It is a civic task that must involve the whole of society. The reasons are obvious: without the necessary social alliances between producers and consumers, the task becomes impossible. In this sense, a change of focus is necessary. Traditionally, Agroecology was excessively focused on mobilizing the food supply, that is, on working with producers, understanding the last link in the chain as a virtually passive final objective, that had to be informed on the benefits of healthy eating but who did not have to be mobilized. This approach has resulted in agroecological experiences multiplying, with the limitations we described above. Producers' decreasing political, economic and even demographic influence explains the meager weight that agrarian policies have on governments' agendas, and those of their supporting political parties. In the middle of the previous decade, Agroecology left the field of agriculture to change focus and look toward the food system, taking all the links of the chain into account in the establishment of a sustainable food strategy (Francis et al., 2003). However, politically speaking, this shift is still not complete: it needs to focus on mobilizing demand or consumption, so that sustainable practices along the entire food chain center on citizens' healthy diet. This is the only way to generate social majorities of change capable of scaling up agroecological experiences and supporting agroecological-oriented local food systems.

The issue of food indeed affects a range of dimensions of social relationships. Satisfying human beings' endosomatic metabolism is increasingly complex, involving aspects related to health, whether physical and mental, well-being of the body, cultural identity, the conservation of material and immaterial heritage, the viability of productive agrarian activities, rural development, the health of agroecosystems, agrifood transformation activities, the sustainability of energy consumption, equity in relations between developed and peripheral countries, etc. Food has become a "thematic meeting point" integrating various social, economic, environmental and political environments, which poses acute, hitherto ignored, governance challenges (Renting & Wiskerke, 2010; Petrini et al., 2016).

The case of Spain is a good illustration of this. The dietary habits of Spaniards are increasingly similar to those of rich countries. Spain consumes a daily average of 3,405 kcal per capita (Schmidhuber, 2006; González de Molina et al., 2017). This diet has led to abandoning healthy Mediterranean habits and adopting others, leading to 41% of the population being overweight (Schmidhuber, 2006, 5). Meat, milk, and other dairy products are directly responsible for this increase. These changes are related to: a growth in per-capita income; the development of supermarkets; changes in food distribution systems; the fact that female workers have less time to cook; growing habits of eating outside the home, often in fast food establishments; as well as a cheapening of livestock products thanks to the low price of labor and raw materials, especially feed imported from third countries (Infante et al., 2018b). The way Spaniards feed themselves has thus undergone a substantial transformation. These changes are one of the major causes of unsustainability, not only in terms of human health but also agroecosystem health, not only of Spaniards but also that of developing countries (UNEP, 2010).[1] Despite the billions of dollars that major brands spend annually on advertising, consumers' concern for the impacts on the environment and health is growing, and both collective and individual mobilization around healthy nutrition is on the rise.

There is, however, another powerful reason that requires consumers to take an active part in the game. It is not merely a question of collective action: organic production that follows agroecological criteria and alternative distribution will not be an effective solution if they are not accompanied by a major change in food consumption patterns and in the values that inspire them. If these patterns do not change, and the intake of meats, eggs, and dairy products are not reduced, even if they are organic, pressures to import food from countries facing food security and hunger problems will intensify. The progress made will be insufficient. Food justice therefore requires a change in the way we meet our endosomatic needs, especially in rich countries. The politicization of food consumption, i.e., the conversion of food into a responsible and therefore a chosen political act is the most effective way to build majorities of change around an alternative food regime and represents one of the populist demands with the biggest capacity to mobilize.

---

[1] Peripheral countries are undergoing the same phenomenon. In Latin America, ultraprocessed food has taken over the diet of children, adults and the elderly. The investigations of the journalist Soledad Barruti (2013, 2018) unveiled disturbing information on intensive and agroindustrial production processes, the poor quality of products offered by big brands, and even on childhood addiction to these types of foods, that in fact are not food.

The most evident way to politicize consumption is to approach human health. Food insecurity has become widespread under the CFR, associated with cases of undernutrition (insufficient food intake to meet the need for food energy), malnutrition (imbalance due to deficient or excessive ingestion of energy and nutrients) and protein–energy malnutrition (resulting from the lack of intake of protein, calories, energy, and macronutrients). Malnutrition is already a widespread phenomenon both in the north and south, related to the increasingly frequent intake of so-called ultraprocessed foods (Monteiro, 2009; Monteiro & Cannon, 2012; Monteiro et al., 2013). In high-income countries, poorest people are the ones to be the most affected by overweight and obesity, because a healthy diet is more expensive than a diet based on processed products, rich in oils, sugars, and fats. A new classification of foods has even been proposed according to processing levels (Monteiro et al., 2010). Dietary guidelines promoted by the CFR are "obesogenic", they do not facilitate the adoption of healthy diets, and they present serious functioning and governance problems that are translating into negative impacts on health at very high costs. They lead to the massive dumping of polluting substances in the soil, air, water courses, and in the food itself.

Another way of politicizing consumption centers on efforts made by many social organizations and even some governmental and parliamentary bodies to recognize a human right to food. This right is defined as "the right to have access, regularly, permanently and freely, either directly, or through cash purchase, to quantitatively and qualitatively adequate and sufficient food, corresponding to the cultural traditions of the population the consumer belongs to, and that guarantees a psychic and physical, individual and collective, satisfactory and dignified life, free of anguish"[2] The right to food is, therefore, a basic and fundamental human right that is in no way guaranteed in view of the worrying problems related to hunger, food insecurity and malnutrition (hidden hunger, overweight, and obesity) that continue to affect swathes of the world's population. Despite being recognized in some international treaties, including the International Covenant on Economic, Social and Cultural Rights (ICESCR), many countries have not yet incorporated the right into their legislation. The right to food is not only a matter of access and enjoyment of a sufficient quantity of food, it is also a matter of nutritional quality and sustainability of its production.

Guaranteeing this right is first and foremost a political issue, a problem of governance, in which the State has a fundamental responsibility but in which society's participation is indispensable. It is essential that the different actors involved in the food system jointly develop public policies. This participation can be channeled by creating spaces to share experiences and generate valid political proposals for all citizens.[3] Food councils (Harper et al., 2009) are a good example of this.

Another way to politicize consumption is represented by the Milan Urban Food Policy Pact (2015)[4]: 160 cities around the world are taking part and governance

---

[2] *Report by the Special Rapporteur on the right to food, Mr. Jean Ziegler, submitted in accordance with Commission on Human Rights resolution 2000/10*, CDESC (E/CN.4/2001/53), February 7, 2001, p. 9.
[3] Significantly, one of the first acts of the new Brazilian government of Jair Bolsonaro, in January 2019 was the shut-down of the National Council for Food Security and Nutrition (CONSEA), generating major citizen protests. The government's decision remained unchanged as we were writing this book.
[4] www.milanurbanfoodpolicypact.org/

instruments have been created around it. It is the first international protocol at the municipal level, aimed at developing sustainable food systems. Its strategic action framework includes recommendations to create contexts facilitating effective action, to foster sustainable and nutritious diets, ensure social and economic equity, to promote food production, improve supply and distribution, and limit food waste. Similar but more specifically agroecological initiatives have emerged the world over. In Spain, for example, it is worth mentioning the *Network of Cities for Agroecology*,[5] whose objective is "to create an exchange process of knowledge, experiences and resources on food policies between Spanish cities that includes local social organizations" and "to establish an operational, agile, specific and common structure that facilitates the process of exchange of knowledge, experiences and resources on food policies between Spanish cities". In the same way, urban and peri-urban agriculture is favoring not only the elimination of barriers between the countryside and the city, but also the politicizing of food consumption in urban areas.

6.7 AGROECOLOGICAL MOVEMENTS AS "NEW GREEN" MOVEMENTS

It has been argued that two traditions scholarly oppose each other regarding studies on the peasantry and the agrarian question: one is based on class analysis, obviously linked to Marxist and neo-Marxist developments, and another is referred to by some as Chayanovian and by others as Neopopulist (White in Bernstein et al., 2018, 706). Our proposal of food populism clearly inscribes itself in this latter stream. Henry Bernstein (2014), as a representative of the first tradition, criticized the vindication of food sovereignty proposed by Vía Campesina, calling it "agrarian populism". Although our proposal of food neopopulism is different, it agrees with Vía Campesina on the importance of recovering food sovereignty from the corporations that govern the CFR on the one hand, and on grounding an alternative regime in the peasantry, without whose mobilization it would not be possible to achieve it, on the other. For Bernstein, this "populist" claim is based on an idealistic conception of the potentialities of the "peasant way" and its presumed ecological virtues. He questions an understanding of the peasantry as a unitary and virtuous peasantry, because it ignores the class differences within it, the complexity and diversity of agents that are involved upstream and downstream in the food chain, and it overestimates the capacity of a substantial segment of the peasants, i.e., the poorest, to increase productivity and solve world food problems.

His thesis on the peasantry and the peasant community are based on the now obsolete Marxist theory on the peasantry's differentiation and disappearance. His central thesis is that there are no peasants nowadays, but rather *petty commodity producers*, subject to Capitalism through an intense commodification process (Bernstein, 2010) that reinforces the tendency toward *class differentiation of petty commodity production* and the emergence of conflicting interests. For him, "large sections of rural people in today's South, perhaps the majority in most places, are better understood as a particular component of 'working class' rather than 'farmers' in any determinate

---

[5] *Red de Ciudades por la Agroecología*: www.ciudadesagroecologicas.eu/

and useful sense" (2014, 1045). For this reason, he criticizes the idyllic conception of community that underlies the approach of food sovereignty supporters: "At the same time, 'community' usually exemplifies 'strategic essentialism' in FS [Food Sovereignty] discourse, as in populist discourse generally, which obscures the contradictions existing within 'communities'" (Bernstein, 2014, 1046).

In Bernstein's writings, clichés abound on farm size and impossible economies of scale, on impossible technological innovation, doubting the ability of peasants to feed the world. Bernstein's Marxist critique is also conducted in the name of a class struggle, whose protagonists have changed and in the name of a class differentiation in agriculture that has failed to lead to the expected proletarianization. Moreover, in postindustrial societies, it is the proletariat itself that has weakened in quantitative terms and has become highly segmented, both in the developed North and in the periphery of the South. Considering the proletariat or the working class to be at the forefront of social change is no longer tenable.

Our theory of peasantry degradation also explains this process of differentiation without assuming that it leads to proletarianization or disappearance, phenomena disproved in practice. Moreover, it emphasizes the market's degree of autonomy and dependence (Ploeg, 1993) and, therefore, the degree of "peasantness" maintained by each peasant social category, thus, the degree of proximity it shows to industrial production and the extent of its subordinate integration in the CFR. From this perspective, observable differences within the peasantry do not question its essential unity or its central role in the struggle for food sovereignty. Different possible peasantry categories, including simple commodity producers, conserve peasant traits to a greater or lesser degree which can serve as the basis for a populist mobilization against the CFR. All these "peasant" social categories are objectively harmed by the CFR and, therefore, are sensitive to a populist appeal to mobilization. The peasantry's segmentation into diverse situations and interests is thus not omitted in our proposal; the segmentation is simply placed in the background in order to highlight the peasantry's essential opposition to capitalism and the CFR.

Bernstein's criticism of Vía Campesina's concept of Food Sovereignty because it is populist does not seem reasonable either. La Vía's political proposal is more class-based than populist because the subject of the agroecological transition is the peasantry, conceived as a unitary whole or with class profiles. However, the task of achieving Food Sovereignty not only belongs to the peasantry, nor does it affect only the peasantry. This is a simplification not even supported by Vía Campesina itself. The global struggle for food sovereignty must be carried out by a very broad range of social groups found along the entire food chain, both in the developed north and in the southern peripheries, starting with peasants and family farmers themselves. The food populism proposal tries precisely to overcome the fragmentation of interests and social groups existing along the chain, through a program of recovery of democratic capacity to decide (sovereignty) what is produced and what is eaten, based on peasant values and on the agroecological rationality that they represent. Consequently, unlike the Vía Campesina proposal, our proposal is populist because it promotes the formation of a sovereign "people", composed not only of the peasantry, but of a much broader spectrum and diversified, though converging, social groups.

Yet, paradoxically, food populism's implicit commitment to the social majorities does not imply having to abandon the class perspective. Rather, we believe that it represents the only way to incorporate it into our postindustrial world. Food populism is the way to build a global discourse capable of producing social change; because of its agroecological content, this discourse is also a class, anti-capitalist discourse. The only class policy that could have any chance to succeed in our dispersed and fragmented societies, is a policy of the people against the elite, the dominating minority, seeking to recover its sovereign capacity to decide on what is produced and what is consumed. In this sense, the nature of the dissent encouraged by food populism is objectively anti-capitalist, that is, objectively contrary to the reproduction of the CFR. If the content of the populist discourse is class, the discourse itself will be based on class too. For us there is no contradiction between populist politics and class politics: the former conditions the possibility of the latter.

This statement can be more clearly understood from the socioecological standpoint adopted in Chapter 1. In industrial metabolism societies, environmental conflicts and social or class conflicts have been apparently separated thanks to the development of money, private property, and the market (Naredo, 2015). In other words, the supremacy of merchandise (money) has overshadowed the environmental conflict spotlighting a class conflict regarding income distribution (surplus value, salary). The institutional capitalist framework and the growing use of fossil fuels have made it possible to separate social or class conflicts from environmental conflicts. Indeed, inequality causes situations that tend to raise social entropy, for example generating relative poverty, deprivation of goods, and social marginalization, discontent, and social protest, etc. The mechanism used by Capitalism in most countries, especially in "developed" countries, has been to compensate for such an increase in social entropy by importing increasing amounts of energy and materials from the environment, progressively raising its metabolic profile (González de Molina & Toledo, 2014, 228). Increasing exosomatic consumption has become an instrument with which to compensate, by building and setting up new and more expensive dissipative structures, a persistent unjust social order, reducing the internal entropy and simultaneously intensifying external entropy, i.e., transferring it to the environment. A significant number of "class" protests have ultimately accelerated the consumption of energy and materials during the twentieth century, especially in the second half of the century.

In this sense, Political Agroecology promotes, through food populism, a form of mobilization; it designs new institutional arrangements to avoid that the reduction of a food system's metabolic profile be transferred to another territory or leads to growing social entropy in the form of rural poverty, malnutrition, undernutrition, or harm to health. On the contrary, the social conflicts generated by unequal access and distribution to food resources tend to be compensated by increases in biophysical or metabolic entropy. The old dichotomy between social equity in land distribution and agrarian income or the conservation of the environment that has crossed in recent years to the left thus does not make sense from a Political Agroecology standpoint. Agrarian sustainability is not possible without social equity, which is not possible without a sustainable use of natural resources.

The struggle for a sustainable food system thus represents a contemporary form of class conflict between the "new food bourgeoisie" – the big corporations in charge of the CFR – and the "new proletariat", that is, all those whom it exploits and oppresses (peasants, new peasants, women, consumers, etc.). As Petersen (2017) claims, "The conflict between industrial metabolism and organic metabolism in agrifood systems is thus a clear contemporary expression of class struggle. However, this class struggle over agriculture and food assumes specific structural forms insofar as capital and labour are dialectically interconnected, forming an organic whole with nature". Agroecology allows the social and environmental aspects of the food process to merge. In this sense, populist mobilizations demanding an alternative regime illustrate the upsurge of a "new environmentalism" (González de Molina et al., 2016) that reconciles social or class protest with environmental protest, preventing the entropic effects of a protest that would be exclusively a class protest or exclusively an environmental protest.

7 The Role of the State and Public Policies

In the light of the previous chapters, the goal of Political Agroecology is to promote a new food regime grounded in sustainability. For this, a new series of norms must be established to organize and regulate metabolic exchanges of energy and materials of a less-entropic nature in agriculture and throughout the food chain generally. The question is that of *metamorphosing* the institutional framework imposed by the corporate food regime (CFR) through the multiplication and interconnection between agroecological experiences. This process, however, will not be possible without changing today's institutional framework. As we saw in Chapter 4, it is necessary to intervene in the State's political environment, given its ability to impose a macro institutional order favorable to the CFR; otherwise, the transforming capacity of agroecological experiences will be neutralized, that is, conventionalized or encapsulated. As we saw in that chapter, the State is currently at the service of the CFR and is obstructing Agroecology upscaling. It is unlikely that it would be possible to scale up through a cumulative expansion of experiences and purporting to do so only creates frustration.

The State is ultimately responsible for agroecosystems' dynamics and degree of sustainability. Farmers' incomes and the stability of the economic units they embody are determined by the prices they pay and receive, in turn strongly conditioned by the regulations and norms established by national public policies. Agroecological collective action is at its most decisive when it is defined by the State and public policy: democratic public institutions must protect agroecological experiences and favor the massification of agroecological production and consumption. But that is not all. The State and public policies must also reverse the "systemic rejection" automatically triggered by the CFR. The ultimate goal of government action and agroecological public policies is that of switching the direction of rejection in such a way that the CFR's common values, behaviors, practices, and institutions be rejected themselves. It is only feasible to reverse this adverse sensitivity within the filters and warning systems through the cooperation between multilevel collective action (movements and social agents) and the democratic State. No single actor working alone, whether the social movements or the State, can bring about the institutional changes necessary for the new food regime to filter and reject the CFR's own behavior. Public policies are, therefore, an essential instrument to set up a new alternative regime that would sustainably govern the production, transformation, distribution and consumption of food.

This chapter focuses on the significant range of experiences accumulated to date in the field. We will draw conclusions on public policy designs (hereon PPs) that facilitate the dissemination of agroecological practices. This chapter is thus a response to the growing demands of the agroecological movement, whose field of action is reaching ever more beyond farms or communities, including the State's administration. It is divided into the following sections: first, PPs are defined in accordance with the approach adopted in this book; second, short-term experiences, but no less powerful ones, of agroecological PPs in different countries are described and conclusions are drawn; third and last, general criteria to elaborate public policies to scale up agroecological experiences is proposed and a range of particularly successful initiatives are highlighted.

7.1   PUBLIC POLICIES FROM A POLITICAL AGROECOLOGY PERSPECTIVE

As we saw in Chapter 1, institutions are dissipative structures that regulate social entropy (social inequality and its effects) as well as metabolic or physical entropy and the trade-off between them. These include political institutions, endowed with a high degree of complexity and self-reflexivity, working both at the micro (family, community, etc.) and macro (State) levels. The role of these political regulators is to synchronize both poles of metabolic exchange and render them stable. This negentropic function of political power is exercised at the expense of generating its own regulatory entropy, in such a way that the balance between the negentropic function and the generated entropy marks the limits of its validity as an institution. Political power reduces entropy by fostering coordination between different stakeholders (individuals and institutions) involved in social metabolism. In short, the task of political institutions is to control and minimize social and physical entropy through information flows, but also through the management of their own internal entropy (transaction costs, bureaucracy, etc.).

This thermodynamic view of political power as an entropy regulator obliges us to jointly consider the two common meanings of the term "politics": "an art of domination" and "an art of integration". The latter suggests the idea of "governability" (Foucault, 1991), that is, the control and governance of a social group settled in a specific territory. From this perspective, the objective of politics is to provide public goods through collective action (Colomer, 2009). Considering the provision of said public goods is beyond citizens' individual reach, a coordinated effort is required, whether through voluntary or coercive means, or whether via collective action or public government institutions which execute public policies. For example, sustainability is a public good that citizens cannot achieve individually. To achieve it, collective action, public policies or a combination of both is required. The production of public goods, which in contemporary societies is dominated by nation states, is largely in the hands of the State (Giddens, 1987), and led by PPs. This task could be understood from a thermodynamic perspective as the production of dissipative structures that are both physical (stocks of goods and services that dissipate energy and materials) and social (norms, regulations that dissipate information flows).

PPs therefore constitute the State's main instrument of action: they represent the "State in action". This conception of PPs, though, is too statist. Some scholars criticize this assimilation of PPs to governance and suggest considering them as the result of the State's interaction with society, that is, they are regarded as a governance issue and are thus subject to the participation of the different groups making up society (Aguilar, 2007; Hufty, 2008; Scartascini et al., 2009; Hoppe, 2010). PPs results from conflict and cooperation dynamics arising in the constitution of public affairs (Torres-Melo & Santander, 2013, 56). From an agroecological viewpoint, we should go even further and institutionalize this interrelation through organized participation and deliberation mechanisms turning PPs into *co-produced* policies (Aguilar, 2007; Subirats et al., 2012, 68). We will come back to this later.

There are at least three major approaches to the origin and nature of PPs. For some, PPs are the State's response to social demands: the State is understood as a "counter" where these demands are received and addressed – a "pluralist" approach (Subirats et al., 2012). A second approach considers that PPs reflect the interests of the social classes that control the State; in our case, these classes would be the corporations making up the CFR and its services. This conception of PPs understands the State as an instrument of the dominant classes (Marxist and neo-Marxist approach). The third approach, the so-called "neo-institutionalist" approach, emphasizes that power is distributed among different actors and their interactions, and examines the organizations and institutional rules they are framed in. According to this approach, public servants or officials are "captured" by the interest groups with which they have privileged relationships (including the so-called "revolving door" effect). However, there is enough empirical evidence to argue that the State makes decisions and implements PPs that respond to a diversity of interests of the dominant classes or that of large corporations. PPs often have a fair margin of freedom, depending on existing power relations between the defense of corporate or class interests and general interest (Subirats et al., 2012, 21). Public powers thus do have some room to maneuver when making decisions, the roots of which often lie in the politicians' desire to retain their positions and the State's necessity to defend its own interests. The more democratic the governments, PP designs and the implementation processes are, the broader the room to maneuver will be.

In any case, PPs are essential tools that must be used by the agroecological movement to transform the industrial metabolic regime and reverse the untenable dynamics that govern the food system. PPs must be oriented toward producing structural changes or lasting institutions that have systemic impacts on the food regime. As we saw in Chapter 3, the production, distribution, and consumption of food are a major cause of global unsustainability and one of the causes of the ecological crisis. The global food system, hegemonized by the CFR, works analogously to the industrial metabolic regime and is one of its essential components. The food system is responsible for 821 million human beings going hungry, two billion being undernourished, and almost another two billion being malnourished; it is responsible for the fact that NCDs (cancer, cardiovascular diseases, diabetes) have overtaken infectious diseases as the main cause of death; it has brought about serious environmental impacts on agroecosystems, but also on water currents, and it carries the major responsibility for greenhouse gas emissions (IPES-Food, 2016).

Agroecology must contribute to achieving the sustainable development goals (SDG), specifically Goal 2: end hunger, achieve food security, improve nutrition, and promote sustainable agriculture. This goal will also favor the achievement of other goals: Goal 1, reducing rural poverty; Goal 3, guaranteeing a healthy diet; Goal 5, promoting equality in rural areas; Goal 6, managing water resources in a sustainable way; Goal 8, providing quality employment throughout the entire food chain; Goal 10, reducing inequality in and between countries; Goal 12, promoting sustainable food consumption patterns; Goal 13, contributing to the mitigation of climate change; Goal 15, maintaining and even increasing biological diversity. The indispensable condition for this to be possible is by substantially reducing the consumption of energy and materials in the global food system as a whole, that is, its metabolic profile, without, however, generating an increase in social entropy, that is, without jeopardizing the right to food. This can only be achieved by transforming the current food regime and replacing it by another one that follows agroecological principles prefiguring in many agroecological experiences that have proliferated around the world. PPs are an essential instrument to make this possible. Without PPs in favor of Agroecology, it is remarkably difficult for agroecological experiences to reach a size enabling them to overcome systemic the critical threshold (social point) necessary to generalize and institute a new sustainable food regime.

7.2 EXPERIENCES IN PUBLIC POLICIES FAVORING AGROECOLOGY

Experience gained in the design and implementation of PPs favoring Agroecology (hereon PPfA) is relatively recent. It dates back virtually to the beginning of this century and to date, very few countries have put these types of PPs into practice. Practices have widely differed in scope and have been implemented at various municipal, regional, and national scales. Many governments have been obliged to implement public policies in favor, at least nominally, of Agroecology, driven by agroecological movements and generalized concerns for the environment and food security. The FAO's intervention in the field has not only encouraged a growing number of governments to set up these kinds of PPs, it has also put concerns about their impacts on the table as well as the need to systematize the experiences developed until now. Despite this, the literature on PPfA is very scarce and reports of international organizations on this matter are even more limited. We dispose of studies in Latin America and Europe that can help conduct an initial analysis of the PPs developed so far. Conclusions can thus be drawn that help designing new and more efficient policies in favor of the scaling up of agroecological experiences.

A study across several Latin American countries was conducted by the Public Policies Network in Latin America and the Caribbean (PP-AL, 2017; available at www.pp-al.org/es). It may constitute the first generally ambitious and aggregate account we have. After an introductory chapter, it describes specific studies in Argentina, Brazil, Chile, Costa Rica, Cuba, El Salvador, Mexico, Nicaragua, and concludes with a comparative regional analysis. Despite this, the study does not include an assessment of the performance of these policies, owing partly to a lack of suitable information sources and scant experience in this type of work. As one of the study's coordinators argues: "In the case of already well-implemented policies, there are no assessments except for very partial ones, either that of the PLANAPO 1 in Brazil or that of the program of recognition of environmental benefits in Costa Rica. Research in Agroecology is generally either still rather incipient, either highly academic and fragmented, weakly oriented toward responding to the social demands of producers" (Sabourin et al., 2017, 210). Several other studies analyze joint schemes or specific policies carried out on that continent, mostly in Brazil, which we will mention later.

No similar study is available for Europe, but some journal articles account for research carried out also in France, the United Kingdom or Spain (Ajates Gonzalez et al., 2018; Ramos García et al., 2018). The analysis carried out in France and the United Kingdom focuses on public policy intentions in favor of Agroecology found in the latter countries' own legal provisions and approved plans. For its part, the Spanish study is actually a first assessment of the measures included in the *II Plan de Fomento de la Agricultura Ecológica de Andalucía* (Second Plan for the Promotion of Ecological Agriculture in Andalusia, by its Spanish name, the biggest territorial region, located in the south of the country). This plan was designed with an agroecological approach. We also dispose of the news items contained in the report on the awards granted following the 2018 call of the *World Future Policy Award*, dedicated on this occasion to rewarding the best policies for scaling up Agroecology, promoting the transition to food systems and sustainable agriculture and climate resilience (World Future Council Foundation, 2018). The award was jointly organized by the United Nations Food and Agriculture Organization (FAO), the World Future Council (WFC), and IFOAM-Organics International. For this, the best agroecological-oriented policies were preselected, including initiatives from Brazil, Denmark, Ecuador, India, the Philippines, Senegal, the USA, as well as TEEBAgrifood. A total of 51 policies were nominated from 25 countries: six from Africa, 12 from Asia, nine from Europe, 20 from Latin America, one from North America, and three international nominations.

We can draw conclusions from all these studies on the current state of the PPfA, their virtues, their shortcomings, and naturally, use them to design a new generation of PPs that help to scale Agroecology. Later, we will also be able to pinpoint the policies we believe may be most useful in achieving that objective, whether due to their focus, their ability to mobilize public resources or the social movements themselves.

The first thing to do is to question the very agroecological nature of many PPfA. Confusion surrounding their definition and specific content has led many PPs to be considered agroecological, when in reality they are not. The confusion takes place on two complementary planes. On the one hand, there are governments' interpretations of Agroecology and its deriving practices. The definition adopted by Latin American countries is generally closer to that of the agroecological movement: it includes not only agrarian practices, but also food-related practices, i.e., distribution and consumption, in addition to the quest for social, gender, race, and ethnic equity. However, when it comes to determining general PPfA principles, definitions become blurred (Sabourin et al., 2017). For its part, the UK and France study shows that both government plans regard Agroecology as a technique for intensifying agriculture. In the case of the UK, Agroecology is considered as a tool alongside other techniques such as GM, under the wider climate-smart agriculture umbrella (UK), to compete in a post-Brexit international market. In the French case, Agroecology is presented as a new paradigm, but in practice it is a sustainable practice among others, being considered as an additional criterion in the list of requirements for subsidies or compensation measures, a way of greening the CAP. In both cases, agroecological practices are technical solutions suitable for maintaining high yields with a limited impact on the environment and coexisting with different (conventional) farming models (Ajates Gonzalez et al., 2018, 12). As stated by these authors, this agroecological approach is focused "more on the aspects … that can be easily co-opted, applied to individual farms, and that do not require a deep transformation of power relations in the current dominant agricultural regime" (Ajates Gonzalez et al., 2018, 6). This latter approach can be found to a greater or lesser degree in all the PPfAs hitherto referred to. In short, the most technological version of Agroecology, stripped of its social change dimension, can be incorporated without any major difficulty into the PPs in favor of the CFR, as an additional technique to solve the problems of industrial agriculture, but without affecting its core.

On the other hand, Agroecology is largely assimilated to organic agriculture (OA): policies in favor of AO are considered PPfA, not only in the studies mentioned above but also in many FAO or IFOAM reports. Definitions of Agroecology and agroecological practices are highly diverse and often ambiguous. Virtually no definition contains clear criteria distinguishing Agroecology from OA. However, the overlaps between OA and Agroecology cannot always be considered negative. A range of agricultural models coexist under the OA umbrella. Some are closer to agribusiness and conventional agroecosystem management, although agrochemicals are not used, and others are closer to agroecological management. In other words, in the experiences analyzed, OA is a management style that facilitates the agroecological transition, either because promotional PPs create a favorable economic and technical context, or because under the OA denomination or seal, markets that would be riskier through conventional avenues can be accessed. In this sense, PPs promoting OA can be objectively positive for farmers wishing to progress toward the agroecological transition. However, it must remain clear that these PPs are double-edged: on the one hand, they favor adopting practices that can be regarded as agroecological as they protect agroecological production against CFR mercantile and institutional aggressions; but on the other, they push toward conventionalization. This is the case, for example, of Argentina's organic production legislation that, according to its evaluators, has been more oriented toward substituting inputs than toward redesigning agroecosystems (Patrouilleau et al., 2017, 36). All evaluated countries largely shared this identification of Agroecology with OA because the approach that best suits the CFR is OA trade or agribusiness; within that model, organic production becomes a seal of differentiated quality and is thus destined to a consumer segment with high purchasing power. Consequently, any PPfA that fails to carry any institutional change leads to encapsulating Agroecology – with the perverse effects this entails as previously described.

One could argue that the environmental policies developed by Latin American and European governments or policies in favor of family farming have followed a similar course. PPs aiming at greater agricultural sustainability or mitigating the impact of chemical agriculture, resulting from international pressures or market demands, can objectively imply some progress in the agroecological transition and favor the implementation of agroecological management practices. For example, Costa Rica's Land Law, or Cuba's National Program to Combat Desertification and Drought (1990), New Environmental Law (1997) and National Environmental Strategy (1997), which pays farmers who follow soil and forest conservation practices. Also worthy of mention are programs to combat climate change that may include and in fact, *do* include agroecological practices, such as Brazil's ABC Plan,[1] or the policies developed in Chile by the "Ministerial Technical Committee on Climate Change"; the NAMA program, dedicated to the mitigation of climate change in agriculture, or other mechanisms such as the ecological blue flag program (Sabourin et al., 2017, 204).

Many countries in Europe and Latin America also have payment schemes for environmental services that can have this positive effect. Indeed, "many countries have developed economic incentives for the protection of the environment and natural resources, which contribute, directly or indirectly, to fostering Agroecology. The programs of Payments for Environmental Service (PES), green bags for the protection of biodiversity (or energy efficiency in Chile) were first developed in Mexico for water protection and then in Brazil with the *PDA* and *Proambiente* programs for tropical rainforest protection. In Costa Rica, the series of agroenvironmental policies is one of the most developed range of policies in favor of family farming by way of instruments and incentives for the adoption of environmentally friendly practices (RBA[2]). They consist in partial, ex post monetary payments (around 20–30% of additional investment) rewarded for adopting certain practices (living barriers, etc.)" (Sabourin et al., 2017, 203). The same could be said of measures to combat climate change in Brazil, Chile or Costa Rica, which can favor the adoption of agroecological practices.

These PPs, however, are fragmentary or sectoral policies. They only encourage one or a reduced range of agroecological practices and therefore do not adopt a holistic approach. Their impact on the agroecological transition is often uncertain and limited. The policies are disconnected from the production, distribution and consumption of food itself. In fact they aim at greening agriculture: "in this technified version of Agroecology, the focus lies in reconnecting agriculture with the environment, and forgetting the reconnection with food, those who grow it, and those who eat it" (Ajates Gonzalez et al., 2018, 14; Gliessman, 2013; Lamine, 2015). Moreover, agroenvironmental measures sometimes actually support conventional agriculture, such as the agroenvironmental aid given to cotton in Spain (González de Molina, 2009). Under the pretext of reducing the numerous chemical treatments applied to this crop, this latter agroenvironmental measure actually represents an income subsidy helping to maintain one of the most polluting and industrialized crops in

---

[1] Plano ABC (Agricultura de Baixa Emissão de Carbono, in Portuguese), that is, Plan Low Carbon Agriculture.

[2] This acronym in Spanish means "recognition for environmental benefits".

Spanish agriculture. In short, most Latin American countries have organic production regulations and some fairly developed environmental measures that address the agricultural sector, although generally, they are limited. Brazil and Nicaragua are the only countries to have achieved a sufficient degree of institutionalization to implement national plans supporting organic production and Agroecology. Nevertheless, the Agroecology Law in Nicaragua is barely enforced due to a lack of funding and regulatory mechanisms (Fréguin-Gresh, 2017, 191). The same has occurred in Brazil since 2017, following substantial budget cuts to PPfA. As this book goes to press, the National Plan of Agroecology and Organic Production is almost totally paralyzed.

Support policies for family farming are also uncertain. Despite a-priori strengthening the peasant sector, past experiences are not being used to increase peasant autonomy, nor is it clear that they have encouraged the introduction of agroecological practices. For example, PPs supporting family farming exist in Chile, Cuba, Brazil, Argentina, El Salvador, Nicaragua, Costa Rica, and Mexico: practically in all Latin American countries. These policies provide technical assistance, credit, etc., and have created spheres for family farmer negotiation and participation. Yet generally, it does not seem that implementing these measures has advanced the transition in those countries, and one could question whether they have delayed the destruction of the peasantry at all (Sabourin et al., 2017, 204).

The success or effectiveness of environmental policies, policies to support family farming, and even policies that are clearly in favor of Agroecology largely depends on the existence of a social movement that demands them. Indeed, the examined experiences show that where agroecological movements and high levels of organization are present, PPs are more effective, relations with the administration are much narrower, and this results the improved functioning and effectiveness of the PP cycle itself. This can be seen in the case of Cuba, where, under the influence of ANAP, there is a better relationship with the Cuban administration and public policies are also more appropriately executed: "The greatest advances can be observed with the Urban, Suburban and Family Agriculture Program, which has consolidated a system of multi-institutional participation at the municipal level, aiming at contributing to food production self-sufficiency. This program covers all urban areas (cities and towns) throughout the 157 municipalities of the country, where progress has also been made in suburban areas, within a ten km radius around cities and five km around small towns" (Vázquez et al., 2017, 124).

The existence of agroecological movements is not essential for PPfA to be implemented, but experience shows that PPfA do become more numerous and effective. The movements are almost a prerequisite for environmental policies or policies in favor of family farming to make agroecological sense. The case of El Salvador, where the agroecological movement is still weak, is a clear illustration of this (Morán et al., 2017). Opposite examples include the case of Brazil (Petersen & Silveira, 2017; Schmitt et al., 2017) where the agroecological movement, strongly articulated with the peasant movement, has achieved unparalleled progress in PPfA, or the case of Andalusia (González de Molina & Guzmán Casado, 2017), where the implementation of plans to promote organic agriculture can only be understood taking into account a peasant movement with strong historical roots, combined with the environmental movement. The Cuban case is another paradigmatic example in that sense. The current that built around the movement *Campesino a Campesino* and ANAP and the urban agriculture movement, that emerged during the special period, have succeeded in expanding the application of the environmental and food legislation adopted by the Cuban State and increasing its agroecological content (Vázquez et al., 2017).

As we have already argued, PPs also constitute areas of confrontation and should not be simplistically regarded as an instrument of the ruling classes to ensure their domination and manage public affairs for their own benefit. The State has room to maneuver, not only because it has its own interests, but also because domination is a power game where the dominated gain ground through mobilization. For example, French agroecological policy responds to CFR demands and to French politicians to counteract public opinion and social movements that demand another model. The situation could be an opportunity to advance the transition: whether PPs are oriented toward Agroecology or toward conventionalization depends on the strength of the social movements.

7.3   MAIN CONCLUSIONS OF THE ANALYSIS OF PUBLIC POLICIES

Let's focus now on the concrete analysis of the PPfA. Although the policies analyzed contain some measures related to the marketing of food and even recommendations or standards to improve food, most PPfA lack a food system approach and center more around the agricultural sector. In addition, they are mostly designed and directed toward agriculture, identifying Agroecology solely with the production of food and not with the technical-economic management and governance of food systems as a whole. In fact, measures regarding land use or land management, which are decisive for the autonomy and sustainability of agroecosystems, are until now outside the scope of PPfA. Only some measures contemplate elaborating alternative marketing channels or the logistics that they would entail, and few are also the measures taken to achieve a more sustainable diet. PPfA lack, therefore, an integral approach, and this is a notable limitation.

We can thus say that most PPfA have a sectoral focus. Many countries, driven by the popularity of Agroecology and the push of agroecological movements, have decided to include agroecological schemes in their policies, yet the schemes are always localized and sectoral by nature. The case of Chile is paradigmatic (Martínez Torres et al., 2017). "In several countries, public policy instruments that could be mobilized in favor of Agroecology are fragile and dispersed. This appears to be the first difficulty, and one that is commonly observed in all countries, including those having specific public Agroecology or organic production policies (Brazil, Costa Rica, Nicaragua, Cuba)" (Sabourin et al., 2017, 209). Experiences in Europe (France and the UK, Andalusia) show a similar tendency, although not to the same extent.

Moreover, in most of the countries studied, PPfAs are relatively marginal with respect to the range of agricultural or food PPs; therefore, their capacity to influence the CFR or conventional agriculture itself is quite limited, enhanced by the fragmented or segmented approach of the measures and their implementation. For example, in Spain, the agroenvironmental measures that support organic farming provide subsidies to farmers who are registered with a certification body and who can present a certificate of having sold their production under the organic seal, but they do not ask for any other requirements that would guarantee that agroecological management practices have been actually adopted. Another illustration consists of agroenvironmental subsidies aimed at combating land erosion that is compatible with the practice of industrial agriculture (González de Molina, 2009). Competences regarding the same agroecological practices are often dispersed over several ministries, hindering their effective application. The promotion of olive waste composting collides in Spain, for example, with environmental legislation that protects aquifers from nitrate pollution. Most examined PPfA do not adopt a necessary systemic or integrated approach; they are not usually part of a coordinated and structured plan. Consequently, their agroecological effectiveness risks being weak. A PPfA should be structured, exploit synergies and be part of a concerted plan.

Our latter affirmation is clearly demonstrated by the fact that most analyzed PPs are part of the environmental policy. The latter reflects the widespread notion among policy makers that Agroecology is an action and research area within the environmental field, restricting the scope of its specific provisions that hardly go beyond the scope of agriculture. In the same way, they mainly target small producers (family farmers) or urban or peri-urban agriculture. It is very difficult to find measures that, even in the agrarian field, adopt a holistic approach including rural development, the profitability of holdings, a gender approach, etc. The sector's social and economic problems are left out or directly trusted to the market or to rural development policies generally, that States develop in other fields of competence. Finally, most of the experiences studied in Latin America and Europe refer to a national scope and do not include municipal PPs, which perhaps accumulate the greatest number of creative experiences. We will return to this later. In the case of Brazil, however, it is worth highlighting that policies supporting Agroecology and organic production have been implemented in at least 11 states (Sabourin et al., in press).

In short, the PPfAs have had a sectoral focus and a limited impact, failing to significantly alter the hegemony of the CFR. What the study's authors described in the case of Brazil is very significant in this sense, especially because Brazil is the country with the most complete agroecological legislation and the most advanced institutionalization of agroecological policies: "Strikingly, despite the country's innovative capacity to implement policies regarding family agriculture, agribusiness has remained largely dominant, creating significant barriers to more structural advances in the construction of an alternative model to rural development" (Schmitt et al., 2017). The authors of the Argentina study draw a conclusion that could apply to all studies: "we can conclude that certain policies over the last decades have fostered visions and agroecological practices in Argentina, but that this has taken place within an institutional system that lacks tools for the integration of policies, therefore encouraging a dual policy development: on the one hand, some policies promote conventional production (including large-scale monoculture, the use of agrochemicals and the use of GMOs) and on the other hand, we can observe alternative experiences such as the case of policies in favor of Agroecology, which cannot be regarded as strategies for the system's production reconversion as a whole" (Patrouilleau et al., 2017, 39).

Many PPfA analyzed were probably launched with the firm intention of responding to growing social demand. Some of them, far from satisfying it result in an attempt to co-opt Agroecology by the CFR and the States. This is what Ajates Gonzalaez et al. (2018, 6) denounce in their analysis of the PPfA in France and the UK, focused "more on the aspects of Agroecology that can be easily co-opted, applied to individual farms, and that do not require a deep transformation of power relations in the current dominant agricultural regime". These authors aim to show "how policies referring to Agroecology can facilitate or hinder its transformative potential to promote fairer and sustainable agrarian socioecological systems in different geographical and cultural contexts" (Ajates Gonzalez et al., 2018, 2).

However, the policies analyzed have contributed both in Europe and in Latin America to the development of a considerable number of agroecological experiences and important progress toward the transition. In that sense, there are cracks in the PPs implemented by governments concerning the development of innovations and agroecological experiences. As already mentioned, the PPfAs analyzed have been the States' response to the growing thrust of social movements and public opinion, but these policies have allowed, in turn, to broaden experiences. We should wonder why they have not had a more extensive territorial scope or reached a more significant share of food consumption. The answer is closely linked to the way the scaling up goals have been defined – that we shall see next – and the role that the PPs should play in that process. To survive, the current institutional framework – i.e., that imposed by the markedly neoliberal CFR – gives all the power to markets, and, even above and beyond national norms, gives priority to private initiative. This explains why these experiences cannot spread or scale up: PPfA are fragmentary, sectoral, low-budget, and do not have the ability to transform the core of the food system. Moreover, the very persistence of this institutional framework explains why many agroecological experiences fail to endure, are aborted (expulsion effect) or languish (encapsulation effect) in productive and territorial areas that are reduced or eventually conventionalized, that is, pushed toward organically producing input substitutions. Consequently, PPfA should mainly target the creation of an institutional framework that favors the upscaling of Agroecology and, if possible, stops the CFR from reproducing itself.

7.4   PUBLIC POLICIES TO SCALE UP AGROECOLOGY

In view of the experiences described, it is necessary to provide an unequivocal definition of PP goals that respond to a clearly agroecological strategy, based on principles that prevent them from being coopted and that effectively guide public action along the path of agricultural sustainability. As explained earlier, the main objective of PPs in favor of Agroecology should be to reduce the metabolic profile of the food system at global and local levels without increasing social and political entropy. For this, the PPs must: (i) promote more sustainable and resilient forms of productive management of agroecosystems, from peasant production to organic agriculture from an agroecological perspective, according to each country's characteristics; (ii) promote access to land and raise peasants' and small farmers' income so as to reduce pressure on resources and keep the population in the countryside; in that way, the reduction of metabolic entropy will not increase social entropy. In this sense, the definition of maximum limits of private appropriation of the land should be a structuring measure according to the agroecological perspective; (iii) reduce the consumption of energy and materials in the food chain, betting on shorter chains, encouraging fresh and seasonal consumption, reducing packaging, etc.; (iv) adopt a healthy diet with lower meat and dairy content, that reduces the food footprint in third countries; and (v) foster the redistribution of resources on a global scale that mitigates unequal ecological exchanges. Reducing the metabolic profile of rich countries' food system and changing diets would undeniably constitute powerful instruments to attain this goal.

For this, States must recover their sovereignty, currently undermined by large food corporations and financial institutions. The sustainable food regime will not be possible without establishing a new democratic model of food governance that makes decisions about what is eaten, how it is produced, and how it is distributed to citizens. It is the State's role to establish mechanisms of democratic participation and deliberation that render the claim of food sovereignty operational, understood as "the right of the people to produce, distribute and consume healthy food close to their territory, in an ecologically sustainable manner" (see Nyéléni Declaration, 2007, https://nyeleni.org/spip.php?article280).

As we saw in previous sections, the current institutional framework regulates agrifood markets for the benefit of conventional production, the great interests of the input industry, large agroindustrial companies and large distribution chains to the detriment of consumers, their own farmers, the environment, and public health. It is also tasked with creating obstacles, through market norms and regulations, to the proliferation of agroecological experiences, imposing costs that penalize their ability to survive and develop. We must reverse this situation, proposing PPs that change the system of monetary and fiscal incentives enjoyed today by conventional production and consumption. The PPfA strategy must be twofold: (i) create favorable conditions for the progressive transformation of production, distribution and consumption along agroecological criteria; but, at the same time (ii) reduce the socioenvironmental impact of conventional agriculture and food, making their real costs visible. This implies drawing up schemes including PPs of a different nature yet articulated to achieve the proposed objectives. We discuss next some criteria that PPfAs should follow their design stage to their implementation stage.

7.5 AN AGROECOLOGICAL APPROACH TO THE DESIGN AND IMPLEMENTATION OF PUBLIC POLICIES

Formally, public policies follow a four-phase cycle: (i) identifying the problem and including it in the government's agenda; (ii) designing the policy, which includes formulating the objectives and the choice of appropriate instruments to carry it out; (iii) implementing it; and (iv) assessing its outcome. In this book, we have described the distance between what a sustainable food system should be and the reality marked by the predominance of a CFR. Problems arise because of that distance and should be identified. Indeed, it does not seem likely that the CFR itself will identify the problems of unsustainability more than partially, and it is even less probable that it will recognize them as collective problems, incorporate them into the government's agenda and find pertinent solutions. In formal democracies, which are largely "captured" by large food corporations and mass media, social mobilization is the only way to include issues in the agenda or to condition it in a hostile institutional context (Meny & Thoenig, 1992; Aguilar, 2003; Duncan, 2015). The formation of agroecological lobbies and "advocacy coalitions" (Sabatier & Jenkins-Smith, 1993; Majone, 2006; Sabatier & Weible, 2014) based on the movements themselves (including those of consumers) and NGOs, is a form of action/pressure toward that end. For this, a power to influence is required that is hardly achieved by mobilizing minorities. The construction of broad alliances through food populism (see Chapter 6) seems to be the most sensible option to force the government's agenda to include public policies that go openly against the CFR or challenge its hegemony.

The second phase, policy design, implies formulating public policy goals consensually with social agents. In the design phase, not only goals, but also the public intervention instruments and the concrete mediations that must be used are determined. Not all options are necessarily viable in a specific legal framework: for example, strong market regulations or prohibitions are expressly rejected by the dominant legal system that refuses any restriction to market and company freedom. Many agroecological PPs require a prior change in the legal system: for example, communal property management – that clashes with private property – or agrarian reform, as a paradigm of this clash. The use of one instrument or the other depends on the power relations existing in each given situation.

There are four instruments for conducting a PP. The first consists of new regulations that modify citizens' behaviors often accompanied by sanctions. For example, prohibiting the use of a certain food additive or particular herbicide because of its proven relationship with cancer or any other disease, based on public health reasons. These types of public policies require a high degree of social consensus to be implemented or very large social majorities to deal with the economic sectors harmed by the prohibition. Their implementation thus depends on power relations and the State's capacity to enforce them.

The second instrument consists of using fiscal instruments with which to modify citizens' behaviors. The latter constitute dissuasive actions that alter economic and social incentives toward certain actions. In a context of market or transition economies, these are very useful instruments to change the order of citizens' preferences. The incentives can be positive, through subsidies, bonuses, or tax deductions, or negative, through taxes, depending on their objective: to encourage a certain behavior or to discourage it. For example, many European governments have imposed taxes to reduce contamination by pesticides (Böcker & Finger, 2016) or nitrates (Rougoor et al., 2001) in superficial and underground water courses, with fairly positive effects; or, for example, the levy that the French, Mexican and British governments introduced on sugar and that taxes soft drink manufacturers, in an attempt to reduce their consumption and fight obesity (*Público*, Spanish newspaper, April 6, 2018). The exploitation agreements in the framework of the Common Agricultural Policy, which encourage the implementation of certain sustainable agricultural practices or payments for environmental services (PES), are examples of positive incentives that farmers receive when they perform sustainable practices.

The third instrument is that of developing information campaigns that convince citizens of the appropriateness of changing their production or consumption behaviors, similar to anti-tobacco campaigns.

The fourth and last instrument are PPs targeting the direct provision of goods and services. For example, building infrastructures or providing services that cannot be undertaken by individual citizens. These types of actions are essential for massification to the extent that cooperatives or associations conducting agroecological food experiences do not usually have the means to face this sort of investment. Building *food hubs*, for example, is considered key to performing leaps of scale in alternative food consumption, given the lack of infrastructures adapted to shorter and more sustainable marketing channels.

These instruments are rarely presented in isolation, but are rather combined in an action plan and even within a same public policy. For example, the re-planning of the territory necessarily requires implementing an agroecological policy to facilitate the closure of biogeochemical cycles, which in turn requires combining several of the aforementioned instruments: from prohibitions that ensure the maintenance of refuge areas for the auxiliary fauna, to incentives to introduce hedges on farms to increase biodiversity, to the implantation of natural fertilizers banks and manure interchange banks, seed exchange banks, etc. We must remember that, as mentioned, these prohibitions are typical of contexts of favorable power relations (contexts in which social movements or social change are empowered), while persuasion through incentives for example, is more specific to market contexts and partial achievements. The provision of public goods and the development of information campaigns for their part, can be found in very different contexts.

The third phase, in which the policy is implemented, requires materializing the goals of the PP through an action plan or scheme that articulates the use of the different instruments (Barret & Fudge, 1981). The action plans set priorities according to the territories and the concrete actions to be carried out, together with the necessary resources. An agroecological PP must have a territorial approach that facilitates adapting the policy to the territory's environmental and sociocultural conditions to which it applies. It is also a logical consequence of the co-produced nature of the PPs that we have defended, a process of participation and deliberation that should not be limited to the design phase, but also to the implementation and evaluation phases. This requires adopting a bottom-up approach, that is, an approach that makes local agents' involvement as effective as possible. The PPs implementation process must thus be lightly defined, providing agents with room to maneuver when applying them in the territory. In this way, there may be different ways of applying the same PP. Therefore, it is essential that territorial plans that gather and adapt the different PP measures to the reality of the territory be elaborated in a participatory and deliberative fashion. Likewise, broad supportive alliances or social coalitions have a key role in the success of agroecological PPs, given that those affected by the measures contained in the plans will try to create, in turn, negative coalitions (Subirats et al., 2012, 209) to prevent the PP from being implemented or to reduce its impact.

The last phase of the PP cycle, policy evaluation, is the analysis, also based on a participatory and territorial approach, of the outcome of implemented PPs and their impact on the environment and society. The evaluation can be done ex ante – particularly recommended in the field of environmental and gender impact – or ex post, to verify whether PPs have indeed led to advances in the generalization of agroecological experiences or, on the contrary, whether they have advanced conventionalization. For example, checking whether they have succeeded in converting an industrial management of agroecosystems into an organic management following agroecological criteria over a significant stretch of land, turning it into a territorially important alternative to industrial agriculture, and making it possible for agents to consider full substitution possible. In the same way, the evaluation may consist in learning to what extent PPs have brought about a significant percentage of organic food to be consumed in the territory in question. This latter outcome undoubtedly requires assessing whether the growth in consumption has arisen from the establishment of alternative channels or whether, on the contrary, the increase in demand has been met through conventional channels. Indicators need to be developed to measure this: for example, the surface area under agroecological management, the number of farming families involved, the percentage of consumption of organic food of agroecological origin, the number of facilities and fairs and other necessary infrastructures for alternative trade, etc.

To summarize, PPfA should contemplate at least four additional criteria or principles of elaboration that distinguish them from usual approaches. In the first place, as we have seen, the effective participation of citizens and especially those affected by PPs must be guaranteed; moreover, PPs should not be the exclusive competence of the State, but should be co-produced, and this is also a key to their success. Second, the PPFAs should have an openly territorial focus, both in their design and implementation, addressing the edaphoclimatic, social, cultural, and economic specificities of the territories where they are to be applied, thus guaranteeing the maximum effectiveness of PPs. The Brazilian policy to combat drought in semiarid regions shows the collaboration between the agroecological movement and the federal government that made it possible for two programs to be managed locally and territorially: the One Million Rural Cisterns Program (P1MC), created in 2003, and the One Land and Two Waters Program (P1 + 2), created in 2007. These examples thus illustrate how dialogue between organized civil society and the State can take place to design and implement public policies (Petersen, 2017; Schmitt et al., 2017). Third, given that our societies are subject to patriarchal domination and that gender inequality was reinforced by the Capitalist system, it is essential that PPs adopt a gender approach making equality possible and preventing discrimination and violence against women. Finally, agroecological PPs must address the rights of future generations, the unborn, and these must be understood as limits to sovereignty and the ability to decide about resources (Serrano, 2007), as we saw in Chapter 4.

7.6 PUBLIC POLICIES THAT LEAD TO SCALING UP

As we have defended throughout this book, agroecological experiences are initiatives that configure autonomous agroecosystems, reducing and eliminating dependence on external inputs and on markets generally, forming shorter and more equitable distribution chains, promoting more sustainable food consumption and promoting higher levels of agricultural resilience to economic or climatic disturbances. In accordance with these principles, PPfA can be evaluated based on their clear agroecological orientation and, at the same time, their capacity to scale up. In the field of productive management of agroecosystems, it could be assessed whether PPs have been able to: (i) rescue, conserve and enhance the use of indigenous genetic material, adapted to soil and climate conditions; (ii) reduce the consumption of fossil energy, promoting, for example, internal circuits, introducing renewable energies, improving the energy efficiency of production, etc.; (iii) close the biogeochemical cycles at the landscape scale or, at least, favor the closing of the cycles of carbon, water, nitrogen, fostering for example a greater and better integration between agriculture and livestock, etc.; (iv) increase biodiversity in such a way that it is practically unnecessary to use phytosanitary products; (v) create spaces for sociotechnical innovation that design technologies adapted to the management needs of family farms; (vi) contribute to the internalization of the social and environmental costs of industrial agriculture; (vii) raise the income of peasants and family farmers, either by assessing the environmental services they provide, or by means of shorter distribution channels that provide more equitable prices for the food produced; (viii) promote gender equality; and (ix) guarantee access to land and other natural resources; etc.

When evaluating the PPfAs that generate the scaling up, the actions that focus on creating distribution systems, alternative to those of the CFR, must play a key role. In this sense, the effectiveness and impact of PPfAs should be analyzed according to whether they: (i) reduce the exosomatic cost of food; (ii) organize the distribution through producer cooperation, eliminating intermediation as much as possible; (iii) build short marketing and distribution channels; (iv) encourage alternative logistics to that of large distribution chains; (v) facilitate the cooperation and networking between food producers, processors, and distributors, creating the conditions for agroecological districts to emerge and consolidate, i.e., agroecological-oriented local food systems (ALOFOODs); (vi) include women, if they distribute the generated income equitably; etc. In the field of consumption, PPfAs will be effective from a scaling up viewpoint: (i) if they advance a shift toward more sustainable food consumption: less meat and feed-raised livestock products, less off-season products, locally produced foods, etc.; and (ii) if they guarantee the majority of the population with access to food, especially lower-income groups, and that organic foods are no longer confined to high-purchasing-power consumer segments.

Below we highlight PPfA that meet some of the criteria outlined above and can be regarded as instrumental in multiplying agroecological experiences, which is currently their principal mission. We do not intend to assemble a catalog of good PPfA practices or to account for all the implemented PPfA, but to highlight a number of them according to the few available ex-post evaluations it is possible to review. Some of them target a specific scope or objective, while others have a comprehensive nature or adopt a multidimensional approach and impact on a substantial part of the food system.

7.6.1 PROGRAM FOR THE CONSTRUCTION OF CISTERNS, BRAZILIAN SEMIARID REGION

The positive impacts that can be achieved when agroecological movements and the State work together, and the need to strengthen the synergies between the two, were exemplified in a public initiative of great social and geographical scale in the Brazilian semiarid region (Petersen, 2017; Sabourin et al., 2017; Schmitt et al., 2017). Influenced by previous NGO-led experiences, the program adopts a contrasting perspective of historic government initiatives in this field: rather than favoring the construction of large hydraulic works (reservoirs), it sought to adapt to the region's characteristics, i.e., a diffused, decentralized, and diversified demand for water. The outcome was ultimately successful. The initiative was instituted via a program for the construction of cisterns for harvesting and storing rainwater, financed by the government but executed by NGOs with the active participation of beneficiary families. As part of the program, families were trained in water management and agroecological production. The first stage of the program was called the One Million Cisterns program (P1MC – *Programa Um Milhão de Cisternas*). Subsequently, as a complement to the first phase, the One Land and Two Waters (P1+2 – *Programa Uma terra e duas Aguas*) was set up, targeting the development of small-scale water storage infrastructures to irrigate fruit or vegetable groves or to meet the needs of domestic and backyard animals (see *Articulação do Semiárido Brasileiro*, at https://asabrasil.org.br/).

7.6.2   ORGANIC AGRICULTURE PROGRAM IN CUBA

Urban gardens had emerged as a popular response to the food crisis caused by the US embargo. The program "Guidelines for Urban Agriculture Subprograms 2008–2010" (*Lineamientos para los Subprogramas de la Agricultura Urbana* 2008–2010) supported this urban agriculture by implementing 28 measures aimed at ensuring a local supply of inputs (organic fertilizers, seeds, etc.). The management of these gardens is mainly based on agroecological criteria and has led to an integral management of production and that of food distribution to a large extent. These successful urban agriculture experiences have brought about the exploitation of unused urban and peri-urban areas, the creation of over 300,000 jobs and a spectacular increase in the production of vegetables (4.2 million tons in 2006). The program currently covers a national surface area of 12,588.91 km^2 representing 14% of the country's agricultural area (Vázquez et al., 2017, 126–127).

7.6.3   ORGANIC FOODS FOR SOCIAL CONSUMPTION IN ANDALUSIA, SPAIN

The program, launched by the Junta de Andalucía (the regional government) during the school year 2005/2006, consisted of supplying organic food to school children in nurseries, primary schools, and state hospitals, accompanied by an educational and information program for children, teachers, patients, their families, and food workers in these centers. A total of 111 schools and 12,000 children were involved during the 2008–2009 academic year. Two state hospitals were also used as pilot centers. It was the first program of this magnitude to be conducted in Spain and a pioneering one in Europe. The utility of this policy for scaling up lies in its comprehensive approach, covering aspects related to health, education, rural development, and the environment that implicate the entire food chain. The program's positive impact extended to Andalusian society as a whole through information activities and experience dissemination.

The program contributes to rural development, helping small and medium-sized farmers to organize themselves for supplying social consumption centers through shorter marketing circuits and creating local groups of producers. This markedly territorial form of organization helps to develop the logistics producer groups need to carry out for distribution, foment a direct relationship between producers and consumers, and ultimately develop the local markets for organic food. The initiative fostered cooperation among different producer groups, who found themselves prompted to diversify production so they could cater as much as possible to market demands; it managed to bring planning to production, and brought producers to complement each other and agree on common sale prices. The program has also created logistical structures to conserve and distribute food (García Trujillo et al., 2009). An average consumption of organic food of over 55% of the total diet was achieved and the program represented a powerful instrument for organic food dissemination among the Andalusian population. Similar programs have been developed in many other places around the world with decidedly positive results. These include Brazil's National School Feeding Program (*PNAE*), described further below.

7.6.4 BIOFERTILIZATION AND BIOLOGICAL CONTROL INPUT PROGRAMS IN CUBA

With the start of the so-called special period in the early 1990s, the impossibility of acquiring chemical inputs from outside the country forced the Cuban State to promote a policy of substitution of chemical inputs by organic and domestic manufacture. The biodiversity conservation programs that had been established in the 1980s were strengthened in 1993. The programs promoted the replacement of chemical pesticides by biological control agents. These programs, which are now integrated, have a network of Entomophagous and Entomopathogenic Reproduction Centers (*CREE*) and several industrial plants for the production of biopesticides, all located in agricultural production areas, with the mission of producing (or reproducing) organisms (bacteria, fungi, nematodes, insects) for local use by farmers (Vázquez et al., 2017).

In the same line, it is worth mentioning the National Program of Organic Fertilizers and Biofertilizers that has been promoting the local production of these bio-inputs since 1991, initially due to the impossibility of acquiring large quantities of chemical fertilizers, and later, to favor land conservation and improvement. The success of these programs is largely due to the support and demands of the ANAP Peasant Farmer Agroecology Movement. Since its creation in 1997, the movement has managed to bring together over 100,000 peasant families, over a third of the existing peasant units in the country (Machín et al, 2010; Vázquez et al., 2017, 120).

7.6.5 THE NATIONAL POLICY OF TECHNICAL ASSISTANCE AND RURAL EXTENSION, BRAZIL

In 2003, the National Policy of Technical Assistance and Rural Extension (PNATER) was launched, following a participatory process promoted by the Ministry of Agricultural Development (MDA). To advance in the implementation of this policy, the Department of Technical Assistance and Rural Extension of the extinct Ministry implemented an extensive training scheme of Extension Agents. Between

2004 and 2010, over 16,000 extension workers were trained. According to Pacífico (2010), 88% of the extension agents carried out rural extension actions supporting the agroecological transition. Despite a number of setbacks, the policy of technical assistance and rural extension was an important lever for the expansion of agroecological experiences in Brazilian family farming (Caporal, 2014).

7.6.6 Institutional Purchasing, Brazil

Brazilian experiences also include two notable programs of public food procurement that have been key to Agroecology's upscaling: the Family Farming Food Acquisition Program (PAA) and the National School Feeding Program (PNAE). The PAA was created in 2003 as one of actions part of the Zero Hunger Program (CONAB, 2018). The scheme aimed at purchasing food directly and exclusively from family farms to supply hospitals, barracks, prisons, university restaurants, kindergartens, and charity schools, in addition to meeting demand by means of farmers' direct food distribution to low income groups in rural and urban areas, among others. The program was coordinated by the Companhia Nacional de Abastecimento (CONAB), a company linked to the Ministry of Agriculture Livestock and Supply (MAPA). Recently one more action was added to the program: the purchase of local non-transgenic seeds (creole seeds and varieties recommended for the region, or organic seeds) for their distribution to family farmers (Porto, 2016).

In 2017, the resources invested in the Plan amounted to R$124,708,501.88. These resources made it possible to purchase 44,407 tons of food produced by 18,688 family farmers organized in cooperatives or associations that submitted 843 projects. (CONAB, 2018). The program allowed CONAB to pay a surcharge for organic products that was 30% above the minimum price established by the government, thus encouraging the agroecological transition.

For its part, the National School Feeding Program (PNAE), popularly known as the "school snack program", was intended to partially meet students' nutritional needs. According to the State Law, at least 30% of the program's resources should be invested in direct purchases from family farmers, prioritizing agrarian reform settlements, traditional communities, indigenous communities and Quilombola communities. The scheme also established the priority of purchasing organic, agroecological and sociobiodiversity products and the possibility of applying up to 30% above the price of conventional foods in local markets. The program's budget in 2017 was 4.15 billion Reais destined to benefit a total of 41 million students.

7.6.7 Olive "Residues" Composting Program in Andalusia, Spain

Within the framework of the Second Andalusian Plan for Organic Farming (2007–2013), notable efforts were made to use locally sourced fertilizing materials. A remarkable example of this has been the organic oil mill waste composting program (crushed olive pulp) initiated in 2007 by the regional government for the purposes of making organic compost. The program focused on oil mills because olive groves in Andalusia occupy a third of the utilized agricultural area (UAA) (INE 2009) and generate a large amount of waste from milling. In this way, access to quality and low-priced compost was facilitated for organic farmers. This is essential in regions with Mediterranean climate conditions, where biomass production is scarce due to lack of water (Guzmán et al., 2011). Between 2007 and 2009, a total of 32,329 hectares benefitted from this production and the value of the compost generated amounted to €919,614 (CAP, 2012). Considering nitrogen as a benchmark nutrient, the olive pulp compost made by farmers along this period replaced 20,625 tons of imported compost between 2007 and 2009 (using Fertiplus for the comparison, an organic material made in Holland), which, in addition to the environmental benefits derived from the local closure of nutrient cycles, also meant that €3.7 million did not leave the Andalusian agricultural sector (Ramos et al., 2018). The number of oil mills with a composting plant rose from four in 2002 to 41 in 2011, the year in which it began to decline due to lack of support (Pérez Rivero, 2016).

7.6.8   ProHuerta's Program, Argentina

ProHuerta was instrumental in disseminating Agroecology among the inhabitants of cities and the surrounding rural environment. The policy aimed at improving the food security of the most vulnerable populations. Although it cannot, strictly speaking, be regarded as a policy to promote Agroecology, it has supported social innovation and the diffusion of Agroecology itself, as accredited by the results of this program present throughout the country. In 2016, a total of 464,527 functioning groves were supported by the program, 676 fairs were held throughout the country, involving 8,562 producers. In the same year, the program spent around 103 million pesos (around US $6.5 million), in addition to the institutional resources provided by INTA. It has also represented a platform of agroecological experimentation applied to small productive units (family and community groves), generating community and institutional learning (Sabourin et al., 2017, 205–206).

Although part of a municipal public policy, it is worth mentioning the experience developed since 2002 around the urban gardens of the city of Rosario by the Secretariat of Social Promotion. The program emerged in response to Argentina's economic crisis, providing food and employment to the most vulnerable urban families. The plots were organically managed, inputs were produced locally, and the surplus sold directly to consumers. Currently, the program involves more than 1,500 producers who produce food for their families and another 250 who sell their surplus. More than two-thirds of the people involved are women. This pioneering experience has inspired other similar initiatives in many Argentine cities and other Latin American cities (Lattuca, 2011).

7.6.9   The National Policy of Agroecology and Organic Production, Brazil

In 2004, the Secretary of Family Farming (SAF) of the Agrarian Development Ministry (MDA) undertook efforts to establish, for the first time, a National Program of Support for Ecologically Based Agriculture in Family Production Units (SAF/MDA, 2004). The Program centered on two "strategic axes": (i) "Subprogram of support for the agroecological transition process" and (ii) "Subprogram of support for the production, commercialization and consumption of organic food", with a duration of two years. Following the 2011 *Marcha de las Margaridas* (a movement of rural women workers) demanding a National Agroecology Policy and following extensive debates, the *Política Nacional de Agroecologia e Produção Orgánica* (National Policy of Organic Agroecology and Production) was launched in 2012. This national plan was governed by a collegial structure including over 28 institutions, of which 14 were governmental and 14 were social organizations. This gave rise to the *I Plano Nacional de Agroecologia e Produção Orgânica* (PLANAPO; First National Plan of Agroecology and Organic Production), that lasted from 2013 to 2015; it was later followed by the Second PLANAPO. The First Plan had a total budget of 8.8 billion Reais (2.5 billions US $) and was organized around four thematic axes with 125 concrete measures. Despite the bigger budget of the Second PLANAPO, it was not effectively implemented due to the coup d'état in Brazil in 2016 followed by the election of a far-right government in 2018.

Both PLANAPOs notably includes the ECOFORTE Program, which encourages the use of financial resources and existing government actions to promote civil society activities in support of Agroecology and organic production, making it possible to expand and strengthen the production, handling and processing of organic and agroecologically produced food. The main beneficiaries include peasant women and young people, agrarian reform settlers, traditional people and communities and their economic organizations (rural enterprises, cooperatives and associations). Over the 2014–2015 period, the program relied on 25 million Reais worth of investments, which funded 28 territorial projects related to Agroecology, extractivism and organic production networks. Resources were later added to finance another 28 networks, with an investment of 32.6 million Reais. Despite problems detected in the projects' execution, the comprehensive and articulated approach of the Second Plans have led to notable advances in the agroecological transition and have contributed to the strengthening of agroecological experiences.

7.6.10   STATE POLICY ON ORGANIC FARMING (2004) AND ORGANIC MISSION (2010), SIKKIM, INDIA

The Himalaya Sikkim is a small state located in Northeast India with around 608,000 inhabitants. Ten per cent of the land, around 75,000 hectares, is farmed. Today, its entire farmland is certified organic. Organic farming is considered the agricultural system closest to the traditional Sikkimese way of farming, which is rain-fed with low external inputs. As such, Sikkim is an excellent model for other Indian states and countries worldwide to scale up Agroecology. Political commitment began in 2003 and in 2010 the state launched the "Organic Mission", an action plan to implement the policy. In 2015, Sikkim declared itself the first organic state in the world. The plan combines mandatory requirements, such as gradually banning chemical fertilizers and pesticides, with support and incentives. Today, more than 66,000 farming families have joined the Organic Mission plan. A total of 80 per cent of the budget between 2010 and 2014 was used to build the capacity of farmers, rural service providers and certification bodies and to support farmers in acquiring certification. In parallel, measures were undertaken to supply farmers with quality organic seeds, such as the strengthening of local organic seed development and production.

One of the most powerful components of the policy was to couple the gradual phase out of subsidies on synthetic inputs with a conversion strategy. The strategy involved training farmers in producing organic inputs such as compost, vermi-compost, and organic pesticides using local plants. More than 100 villages with 10,000 farmers in all four districts of the state benefited from these programs during the first pilot phase of the mission (2003–2009). The policy is also notable for the holistic approach it adopted: it is a comprehensive policy that tackles many aspects needed for the transition toward organic farming (input provision, capacity building, etc.). This plan, winner of the *Future Policy Award* in its 2018 edition, dedicated to awarding PPs in favor of scaling up Agroecology, confuses in some way organic farming and Agroecology, as do many other plans worldwide. The weakest point is its market-oriented approach to organic farming: organic growing is seen as an excellent strategy to target national and international markets, but not food sovereignty (World Future Council & IFOAM, 2018).

A similar experience, perhaps with more agroecological content, is the Climate Resilient, Zero Budget Natural Farming (ZBNF) Program (2015, Andhra Pradesh, India). ZBNF stands for an Indian agroecological farming movement. It includes methods to eliminate external inputs, restore ecosystem health and build climate resilience through diverse, multi-layered cropping systems. By March 2018, a total of 160,000 farmers in 1,000 villages across all the 13 districts of Andhra Pradesh had started to practice ZBNF. The goal is to reach 500,000 farmers by March 2019. A unique feature is the program's bottom-up approach. Recently, Andhra Pradesh adopted the overall vision to become a natural farming state.

7.6.11   MILAN URBAN FOOD POLICY PACT (2015)

This is the first global protocol on food at the municipal level. As of July 2018, the Pact was signed by mayors of 171 cities worldwide, representing 450 million inhabitants. Cities commit to developing sustainable food systems that are inclusive, resilient, safe, and diverse, that provide healthy and affordable food to all people within a human rights-based framework. Cities strive to minimize waste and conserve biodiversity while adapting to and mitigating the impacts of climate change. The Pact recommends 37 specific actions that cover the entire scope of the food system.

7.6.12   ADVANCES IN PROFESSIONAL TRAINING AND SUPPORT FOR THE ORGANIZATION OF AGROECOLOGY HUBS IN BRAZILIAN EDUCATIONAL INSTITUTIONS

Although the policies are not directly aimed at scaling up Agroecology, they are indirectly making a major contribution to the construction of agroecological knowledge and the implementation of concrete experiences in the field. In the field of formal education, the Ministry of Education (MEC) approved the inclusion of Agroecology Training in its curricula of medium- and higher-level degrees, instituting professionalization in this area. According to Massukado and Balla (2016), "Currently, 33 higher education study programs in Agroecology are running in Brazil, offered by 22 higher education institutions, some with Agroecology programs in more than one campus. Of the total number of higher-education level study programs, 27 are technological (82%) and 6 are academic (18%), with an annual offer of approximately 1,700 places". At the graduate level, there are 31 specialization courses (360 hours) offering 1,500 places per year. In addition, there are eight Master's degrees and one Doctorate degree. The same authors found in their research that between 2008 and 2016, the number of research groups in Agroecology registered in the National Council of Science and Technology (CNPq) went from 101 to 381 groups and the number of researchers linked to these groups rose from 2,383 to 12,277 in 2016, of which 3,819 were doctors (Massukado & Balla, 2016).

With the financial support of the Ministry of Agriculture and the Ministry of Agrarian Development, the National Conselho de Ciência e Tecnologia (CNPq) promoted the creation of Agroecology Hubs in higher and medium-education institutions by means of calls for projects. The result was the constitution of 115 centers, with the participation of 437 teachers and 1,582 students (Ferreira, 2016). According to this author, between 2012 and 2016, the Hubs taught Agroecology courses to 25,530 students from different degrees and to 6,372 rural extension agents; they published 1,049 articles and carried out 1,460 seminars/meetings. This policy of support to Agroecology training hubs is undoubtedly the largest initiative in the world. There are currently around 220 hubs in Brazilian educational institutions and they undeniably constitute a powerful driver of the agroecological transition.

References

Adams, R.N. 1975. *Energy and Structure. A Theory of Social Power*. Austin: University of Texas Press.

Adams, R.N. 1988. *The Eight Day: Social Evolution as the Self-Organization of Energy*. Austin: University of Texas Press.

Adams R.N. 2001. *El octavo día*. Mexico: Universidad Autónoma Metropolitana.

AEAT. 2014. Datacomex: Estadísticas del comercio exterior español. Agencia Española de Administración Tributaria. http://datacomex.comercio.es

Aganbem, G. 2006. *El reino y la Gloria. Por una genealogía teológica de la economía y del gobierno*. Valencia: Editorial Pretextos.

Agarwal, B. 2010. *Gender and Green Governance: The Political Economy of Women's Presence Within and Beyond Community Forestry*. Oxford: Oxford University Press.

Agnoletti, M. (ed.). 2006. *The Conservation of Cultural Landscapes*. Wallingford/Cambridge MA: CAB International.

Aguilar, L. 2003. Estudio introductorio. In L.F. Aguilar (ed.). *Problemas políticos y Agenda de Gobierno*. Mexico: Grupo Editorial Miguel Angel Porrua.

Aguilar, L. 2007. El aporte de la Política Pública y de la Nueva Gestión Pública a la gobernanza Revista del CLAD. *Reforma y Democracia*, 39, 1–15.

Ajates Gonzalez, R., Thomas, J., & Chang, M. 2018. Translating agroecology into policy: the case of France and the United Kingdom. *Sustainability*, 10, 2930; doi:10.3390/su10082930.

Alcott, B. 2005. Jevons paradox. *Ecological Economics*, 54, 9–21.

Alesina, A., Algan, Y., Cahuc, P., & Giuliano, P. 2015. Family values and the regulation of labor. *Journal of the European Economic Association*, 13, 599–630.

Allen, R.C. 2004. *Revolución en los campos. La reinterpretación de la revolución agrícola inglesa*. Zaragoza: SEHA/ Prensas Universitarias de Zaragoza.

Allen, P., & Kovach, M. 2000. The capitalist composition of organic farming: the potential of markets in fulfilling the promise of organic farming. *Agriculture and Human Values*, 17, 221–232.

Altieri, M. & Toledo, V. 2011. The agroecological revolution in Latin America. Rescuing nature, ensuring food sovereignity and empowering peasant. *Journal of Pesant Studies*, 38, 587–612.

Altieri, M.A., & Nicholls, C.I. 2007. *Biodiversidad y manejo de plagas en agro sistemas*. Barcelona: Icaria.

Altieri, M., Nicholls, C.I., & Funes-Monzote, F. 2012. The scaling up of Agroecology: spreading the hope for food sovereignty and resiliency: A contribution to discussions at Rio+20 on issues at the interface of hunger, agriculture, environment and social justice. Socla. https://foodfirst.org/wp-content/uploads/2014/06/JA11-The-Scaling-Up-of-Agroecology-Altieri.pdf.

Amore, M.D., & Epure, M. 2018. Family ownership and trust during a financial crisis (May 15, 2018). Available at SSRN: https://ssrn.com/abstract=2968889 or http://dx.doi.org/10.2139/ssrn.2968889

Anisi, D. 1992. *Mercado, valores: una reflexión económica sobre el poder*. Madrid: Alianza Editorial.

Arrow, K.J. 1974. *Elección social y valores individuales*. Ministerio de Economía y Hacienda.

Axelrod, R. 2004. *La complejidad de la cooperación: modelos de cooperación y colaboración basados en los agentes*, 2ª edición. Mexico: Fondo de Cultura Económica.

Axelrod, R. 2006. *The Evolution of Cooperation*, rev. ed. New York: Perseus Books.

Bailey, K.D. 1990. *Theory of Social Entropy*. New York: SUNY Press.

Bailey, K.D. 1997a. The autopoiesis of social systems: assessing Lymahnn's Theory of selfreference. *Systems Research and Behavioral Science*, 14, 83–100.

Bailey, K.D. 1997b. System entropy analysis. *Kybernetes*, 26, 674–688.

Bailey, K.D. 2006a. Living systems theory and social entropy theory. *Systems Research and Behavioral Science*, 23, 291–300.

Bailey, K.D. 2006b. Sociocybernetics and social entropy theory. *Kybernetes*, 35, 375–384.

Barkin, D., Batt, R., & DeWalt, B.R. 1991. *Alimentos versus Forrajes: la sustitución entre granos a escala mundial*. Madrid: Siglo XXI Editores.

Barrera-Bassolls, N. & Toledo, V.M. 2008. *La memoria biocultural. La importancia ecológica de las sabidurías tradicionales*. Barcelona: Icaria.

Barret, S., & Fudge, C. 1981. *Policy and Action: Essays on the Implementation of Public Policy*. London and New York: Methuen.

Barruti, S. 2013. *Mal Comidos: cómo la industria alimentaria argentina nos está matando*. Ciudad Autónoma de Buenos Aires: Planeta.

Barruti, S. 2018. *Mala Leche: el supermercado como emboscada – Por qué la comida ultraprocesada nos enferma desde chicos*. Ciudad Autónoma de Buenos Aires: Planeta.

Barton, R.A., & Harvey, P.C. 2000. Mosaic evolution of brain structure in mammals. *Nature*, 405(6790), 1055–1058.

Beck, U. 1998. *La sociedad del riesgo: hacia una nueva modernidad*. Barcelona: Paidós.

Bénabou, R., & Tirole, J. 2003. Intrinsic and extrinsic motivation. *The Review of Economic Studies*, 70, 489–520.

Benyus, J.M. 1997. *Biomimicry: Innovation Inspired by Nature*. New York: William Morrow, USA.

Bergh, J.C. van der, & Bruinsma, F.R. (eds.). 2008. *Managing the Transition to Renewable Energy*. Cheltenham: Edward Elgar.

Bermudez Gómez, C.A. 2011. Mercosur y Unasur: una mirada a la integración regional a comienzo del siglo XXI. *Análisis Político*, 72, 115–131.

Bernstein, H. 1977. Notes on capital and peasantry. Review of African Political Economy. Vol. 10:60–73.

Bernstein, H. 1986. Capitalism and petty commodity production. *Social Analysis: Journal of Social and Cultural Practice*, 20, 11–28.

Bernstein, H. 2001. The peasantry in global capitalism. In L. Panitch & C. Leys (eds.). *Socialist Register, Working Classes, Global Realities*. New York: Monthly Review Press.

Bernstein, H. 2010. *Class Dynamics of Agrarian Change, Agrarian Change and Peasant Studies Series*. Halifax, N.S., Sterling, VA: Fernwood Publishers, Kumarian Press.

Bernstein, H. 2014. Food sovereignty via the 'peasant way': a sceptical view. *The Journal of Peasant Studies*, 41, 1031–1063.

Bernstein, H., Friedmann, H., van der Ploeg, J.D., Shanin, T., & White, B. 2018. Forum: Fifty years of debate on peasantries, 1966–2016. *Journal of Peasant Studies*, 45, 689–714.

Bicchieri, C. 2016. *Norms in the Wild. How to Diagnose, Measure, and Change Social Norms*. Oxford: Oxford University Press.

Biovision. 2018. Beacons of Hope: Path to a more sustainable food system. www.biovision.ch/en/news/beacons-of-hope-path-to-a-more-sustainable-food-system/ (accessed January 4, 2019).

Blaikie, P. 2008. Epilogue: Towards a future for political ecology that works. *Geoforum*, 29, 765–772.

Blaikie, P. & Brookfield H. 1987. *Land Degradation and Society*. London: Methuen.

Böcker, T. & Finger, R. 2016. European pesticide tax schemes in comparison: an analysis of experiences and developments. *Sustainability*, 8, 378.

Boltansky, L. & Chiapello, E. 2002. *El nuevo espíritu del capitalism* [*The New Spirit of Capitalism*]. Madrid: Editorial Akal.

Boltzmann, L. 1964. [1896]. *Lectures on Gas Theory*. Berkeley: University of California Press.

Boulding, K.E. 1978. *Ecodynamics: A New Theory of Societal Evolution*. London: Sage Publications.

Boulding, K.E. 1994 [1992]. Kenneth E. Boulding. De la química a la economía y más allá. In M. Szenberg (ed. 1994). *Grandes economistas de hoy*. Madrid, Debate, pp. 79–95.

Bourdieu, P. 1979. *La distinction. Critique sociale du jugement*. Paris: Les Éditions de Minui.

Bourdieu, P. 1991. El sentido práctico. Madrid: Editorial Taurus.

Bourdieu, P. 2004. El baile de los solteros: la crisis de la sociedad campesina en el Bearne. Barcelona: Editoral Anagrama.

Bowen, S. 2010. Embedding local places in global spaces: geographical indications as a territorial development strategy. *Rural Sociology*, 75, 209–243.

Bowen, S., & De Master, K. 2011. New rural livelihoods or museums of production? Quality food initiatives in practice. *Journal of Rural Studies*, 27, 73–82.

Braudel, F. 1995. *Civilização material, economia e capitalismo séculos xv-xviii*. São Paulo: Martins Fontes.

Brenner, R. 2009. *La economía de la turbulencia global*. Madrid: Akal.

Brescia, S. (org.). 2017. *Fertile Ground; Scaling Agroecology from the Ground Up*. Oakland: Food First Books.

Bringenzu, S. 2015. Possible target corridor for sustainable use of global material. Resources, 41, 25–54.

Bronley, D.W. 2016. *Sufficient Reason: Volitional Pragmatism and the Meaning of Economic Institutions*. Princeton: Princeton University Press.

Bruckner M, Giljum S, Lutz C, Wiebe KS. 2012. Materials embodied in international trade-global material extraction and consumption between 1995 and 2005. *Glob. Environ. Change* 22(3):568–76.

Brunori, G., Rossi, A., & Guidi, F. 2012. On the new social relations around and beyond food. Analysing consumers' role and action in Gruppi Di Acquisto Solidale (solidarity purchasing groups). *Sociologia Ruralis*, 52, 1–30. doi:10.1111/j.1467-9523.2011.00552.x.

Bruton, H.J. 1998. A reconsideration of import substitution. *Journal of Economic Literature*, 36, 903–936.

Buckwell, A., Nordang Uhre, A., Williams, A., Jana Poláková, J., Blum, W.E., Schiefer, J., Lair, G.J., Heissenhuber, A., Schiel, P., Krämer, C., et al. 2015. *The Sustainable Intensification of European Agriculture*. Brussels: RISE Foundation. Available online: www.risefoundation.eu/images/files/2014/2014_SI_Brief.pdf (accessed October 30, 2015).

Buggle, J., & Durante, R. 2017. *Climate Risk, Cooperation, and the Co-Evolution of Culture and Institutions*. London: Centre for Economic Policy Research.

Bui, S., Cardona, A., Lamine, C., & Cerf, M. 2016. Sustainability transitions: insights on processes of niche–regime interaction and regime reconfiguration in agri-food systems. *Journal of Rural Studies*, 48, 92–103.

Bulatkin, G.A. 2012. Analysis of energy flows in agro-ecosystems. *Herald of the Russian Academy of Sciences*, 82, 326–334.

Bunge, M. 2015. *Emergencia y convergencia*. Cham: Springer Nature.

Cadieux, K.V., Carpenter, S., Blumberg, R., Liebman, A., & Upadhyay, B. 2019. Reparation ecologies: regimes of repair in populist agroecology. *Annals of the American Association of Geographers*. Available at: http://works.bepress.com/kvalentine-cadieux/34/

Calva, J.L. 1988. *Los Campesinos ante el devenir*. Mexico: Siglo XXI Editores.

Campbell, B.M., Thornton, P., Zougmore, R., van Asten, P., & Lipper, L. 2014. Sustainable intensification: what is its role in climate smart agriculture? *Current Opinion in Environmental Sustainability*, 8, 39–43.

Canning, P., Charles, A., Huang, S., Polenske, K., & Waters, A. 2010. Energy use in the U.S. Food System. Economic Research Report 94. Washington, DC: United States Department of Agriculture.

CAP (Consejería de Agricultura y Pesca de la Junta de Andalucía). 2007. *II Plan andaluz de Agricultura Ecológica*. Sevilla: Consejería de Agricultura y Pesca.

CAP (Consejería de Agricultura y Pesca de la Junta de Andalucía). 2012. Boletín de compostaje para la producción ecológica. Primer trimestres 2012. www.juntadeandalucia.es (accessed March 27, 2014).

Caporal, F.R. 2014. Extensão Rural como Política Pública: a difícil tarefa de avaliar. In Sambuichi, R.H.R., Silva, A.P.M., Oliveira, M.A.C., & Savian, M. (orgs.) *Políticas Agroambientais e Sustentabilidade: desafios, oportunidades e lições aprendidas*. Brasília: IPEA, pp. 19–47.

Carneiro, F.F, Rigotto, R.M., Augusto, L.G.S., Friedrich, K. & Búrigo, A.C. (orgs.). 2015. *Dossié Abrasco: um alerta sobre os impactos dos agrotóxicos na saúde*. Rio de Janeiro: EPSJV; São Paulo: Expressão Popular.

Centola, D., Becker, J., Brackbiill, D., & Baronchelli, A. 2018. Experimental evidence for tipping points in social convent. *Science*, 360(6393), 1116–1119.

CEPAL, FAO, IICA. 2012. *Perspectivas de la agricultura y del desarrollo rural en las Américas: una Mirada hacia América Latina y el Caribe*. Santiago, Chile: FAO.

Charão Marques, F., Ploeg, J. D. van der, Kessler Dal Soglio, F., Barbier, M., & Elzen, B. 2012. New identities, new commitments: something is lacking between niche and regime. In *System Innovations, Knowledge Regimes, and Design Practices towards Transitions for Sustainable Agriculture*. Paris: INRA – Science for Action and Development, pp. 23–46.

Chayanov, A. 1966a. On the theory of non-capitalist economic system. In D. Thorner, B.H. Kerblay, & R.E.F. Smith (orgs.), *A.V. Chayanov on the Theory of Peasant Economy*. Madison: The University of Wisconsin Press.

Chayanov, A 1966b. The theory of peasant economy. In D. Thorner, B.H. Kerblay, & R.E.F. Smith (orgs.), *A.V. Chayanov on the Theory of Peasant Economy*. Madison: The University of Wisconsin Press.

Chevassus-au-Louis, B. & Griffon, M. 2008. *La Nouvelle Modernité: Une Agriculture Productive à Haute Valeur Ecologique*. Available online: http://clubdemeter.com/pdf/ledemeter/2008/la_nouvelle_modernite_une_agriculture_productive_a_haute_valeur_ecologique.pdf (accessed September 1, 2016).

Cleveland, C., 1995. The direct and indirect use of fossil fuels and electricity in USA agriculture, 1910–1990. *Agriculture, Ecosystems & Environment*, 55, 111–121.

Clune, S., Crossin, E. & Verghese, K. 2017. Systematic review of greenhouse gas emissions for different fresh food categories. Journal of Cleaner Products, 140, 766–783.

Clunies-Ross, T., & Hildyard, N. 2013. *The Politics of Industrial Agriculture*. London: Routledge.

Coase, R.H. 1994. *La empresa, el mercado y la ley*. Madrid: Alianza Editorial.

Colomer, J.M. 2009. *Ciencia de la política: Una introducción*. Barcelona: Editorial Ariel.

CONAB-Companhia Nacional de Abastecimento. 2018. Compêndio de Estudos Conab. *Programa de Aquisição de Alimentos* (PAA: Resultado das ações da CONAB em 2017). Brasília: CONAB.

Costanza, R., de Groot, R., Sutton, P.C., van der Ploeg, S., Anderson, S., Kubiszewski, I., Farber, S., & Turner, R.K. 2014. Changes in the global value of ecosystem services. *Global Environmental Change*, 26, 152–158.

Costanza, R., Graumlich, L., Steffen, W., Crumley, C., Dearing J., Hibbard, K., Leemans, R., Redman, C., & Schimel, D. 2007. Sustainability or collapse: what can we learn from integrating the history of humans and the rest of nature? *AMBIO: A Journal of the Human Environment*, 36, 522–527.

Costanza, R., Graumlich, L.J. & Steffen, W. 2005. *Sustainability or Collapse? An Integrated History and Future of People on Earth*. Cambridge, MA: The MIT Press.

CSM (Food Security and Nutrition Civil Society Mechanism). 2018. *Connecting smallholdings to markets; an analytical guide*. Rome: CSM.

Cuesta, P. & Gutiérrez, P. 2010. El equipamiento comercial de los centros comerciales en España. *Distribución y Consumo*, 110 (Marzo–Abril), 121.

Cunfer, G. & Krausmann, F. 2009. Sustaining soil fertility: agricultural practice in the Old and New Worlds. *Global Environment*, 4, 9–43.

Cussó, X., Garrabou, R., Olarieta, J.R., & Tello, E. 2006. Balances energéticos y usos del suelo en la agricultura catalana: una comparación entre mediados del siglo XIX y finales del siglo XX. *Historia Agraria*, 40, 471–500.

Daly, H. 1973. *Toward A Steady Estate Econmy*. San Francisco: W.H. Freeman.

Darnhofer, I.S. 2014. Contributing to a transition to sustainability of agri-food systems: potentials and pitfalls for organic farming. In S. Bellon, & S. Penvern (eds.). *Organic Farming, Prototype for Sustainable Agricultures*. Dordrecht: Springer Science, pp. 439–452.

Darnhofer, I. 2015. Socio-technical transitions in farming. Key concepts. In L.A. Sutherland, I. Darnhofer, G. Wilson, & L. Zagata (eds.). *Transition Pathways towards Sustainability in Agriculture. Case studies from Europe*. Wallingford: CABI.

Darnhofer, I., Lindenthal, T., Bartel-Kratochvil, R., & Zollistsch, W. 2010. Conventionalisation of organic farming practices: from structural criteria towards an assessment based on organic principles. A review. *Agronomy for Sustainable Development*, 30, 67–81.

De Schutter, O. 2011. *Agroecology and the Right to Food*. ONU (report presented at the 16th Session of the United Nations Human Rights Council).

Deleuze, G. & Guattari, F. 1995. *Mil platôs; capitalismo e esquizofrenia* (V1). São Paulo: Editora 34.

Dennet, D. 1996. *Contenido y conciencia*. Barcelona: Gedisa.

Diamond, J. 2004. *Colapsos*. Madrid: Editorial Debate.

Dittrich, M. & Bringezu, S. 2010. The physical dimension of international trade. Part 1: Direct global flows between 1962 and 2005. *Ecological Economics*, 69, 1838–1847.

Dittrich, M., Bringezu, S., & Schütz, H. 2011. The physical dimension of international trade, part 2: Indirect global resource flows between 1962 and 2005. *Ecological Economics*, 79, 32–43.

Dixon, J., Gulliver, A., & Gibbon, D. 2001. *Sistemas de producción agropecuaria y pobreza. Cómo mejorar los medios de subsistencia de los pequeños agricultores en un mundo cambiante*. Roma: FAO.

Domenech, A. 1998. Ocho desiderata metodológicos de las teorías sociales normativas. *Isegoría*, 18, 115–141.

Dörr, F. 2018. Food regimes, corporate concentration and its implications for decent work. In C. Scherrer, & S. Verma (eds.). *Decent Work Deficits in Southern Agriculture: Measurements, Drivers and Strategies*. Labor and Globalization Volume 11. Kassel: Rainer Hampp Verlag, 178–208.

Duncan, J. 2015. *Global Food Security Governance: Civil Society Participation in Committee on World Food Security*. London: EarthScan Routledge.

Echevarría, C. 1998. A three-factor agricultural production function: the case of Canada. *International Economics Journal*, 2, 63–76.

Edwards-Jones, G., Milà I Canals, Ll., Hounsome, N., Truninger, M., Koerber, G., Hounsome, B., Cross, P., York, E.H., Hospido, A., Plassmann, K., et al. 2008. Testing the assertion that 'local food is best': the challenges of an evidence-based approach. *Trends in Food Science & Technology*, 19, 265–274.

Egea-Fernández, J.M., & Egea-Sánchez, J.M. 2012. Canales cortos de comercialización, soberanía alimentaria y conservación de la agrobiodiversidad. In *Actas del X Congreso de Agricultura Ecológica*. Albacete: SEAE.

Eisenmenger, N., Ramos, J., & Schandl, H. 2007. Transition in a contemporary context: patterns of development in a globalizing world. In M. Fisher-Kowalski & H. Haberl (eds.). *Socioecological Transitions and Global Change. Trajectories of Social Metabolism and Land Use*. Cheltenham: Edward Elgar, pp. 179–222.

Elster, J. 2007. *Explaining Social Behavior: More Nuts and Bolts for the Social Sciences*. Cambridge: Cambridge University Press.

Elzen, B., Van Mierlo, B., & Leeuwis, C. 2012. Anchoring of innovations: assessing Dutch efforts to harvest energy from glasshouses. *Environmental Innovation and Societal Transitions*, 5, 1–18.

Engel, G.S., Calhoun, T.R., Read, E.L., Ahn, T.-K., Mančal, T., Cheng, Y.-C., Blankenship, R., & Fleming, G.R. 2007. Evidence for wavelike energy transfer through quantum coherence in photosynthetic systems. *Nature*, 446(7137), 782–786.

EU-DG AGRI (European Commission. Directorate-General For Agriculture And Rural Development). 2010. *An Analysis of the EU Organic Sector*. Brussels: European Commission.

FAO. 1995. Agricultural trade. Entering a new era? www.fao.org/es/esa/es/pubs_sofa.htm.

FAO. 2000. *El estado de la inseguridad alimentaria en el mundo*. Rome: FAO.

FAO. 2004. *FAO Statistical Yearbook 2004*. Rome: FAO.

FAO. 2006. Livestock's Long Shadow: Environmental Issues and Options. November 29. www.fao.org/3/aa0701e.pdf

FAO. 2007. *SOFA (The State of Food and Agricultura [El estado mundial de la agricultura y la alimentación]*. Rome: FAO.

FAO. 2008. *Current World Fertilizar Trenes and Outlook to 2011/12*. Rome: FAO.

FAO. 2009. Global Agriculture towards 2050. Report from the High-Level Expert Forum 'How to Feed the World 2050'. Available online: www.fao.org/fileadmin/templates/wsfs/docs/Issues_papers/HLEF2050_Global_Agriculture.pdf (accessed May 4, 2014).

FAO. 2011. *State of Land and Water*. Rome: FAO.

FAO. 2012a. *Dinámicas del Mercado de tierra en América Latina y el Caribe: concentración y extranjerización*. Santiago, Chile: FAO.

FAO. 2012b. *Directrices sobre la Prevención y Manejo de la Resistencia a los Plaguicidas*. Rome: FAO.

FAO. 2016. *El estado mundial de la agricultura y la alimentación*. Rome: FAO.

FAO. 2018. Scaling *up Agroecology Initiative: Transforming Food and Agricultural Systems in Support of the SDGs*. Rome: FAO.

Fath, B.D., Jørgensen, S.E., Patten, B.C., & Străskraba, M. 2004. Ecosystem growth and development. *Biosystems*, 77, 213–228.

Federici, S. 2018. *El patriarcado del salario. Critica feminista al marxismo*. Madrid: Traficantes de sueños.

FEHR. 2005. *Diagnóstico del sector de bares, restaurantes y cafeterías*. Madrid: Federación Española de Hostelería.

Ferrajoli, L. 2010. *Democracia y garantismo*. Valencia: Trotta.

Ferreira, T.L. 2016. Sistematização dos Impactos das Chamadas 46/2012 e 81/2013 (MCTI, MAPA, MEC, MDA. Apresentação em PDF. BLOG: frcaporal.blogspot.com.br (accessed July 16, 2018).

Fischer-Kowalski, M. 2011. Analyzing sustainability transitions as a shift between socio-metabolic regimes. *Environmental Innovation and Societal Transitions*, 1, 152–159.

Fischer-Kowalski, M., & Amann, C. 2001. Beyond IPATS and Kuznets curves: globalization as a vital factor in analysing the environmental impact of socioeconomic metabolism. *Population and Environment* 23, 7–47.

Fischer-Kowalski, M., & Haberl, H. (eds.). 2007. *Socioecological Transitions and Global Change. Trajectories of Social Metabolism and Land Use*. Cheltenham: Edward Elgar.

Fischer-Kowalski, M., & Rotmans, J. 2009. Conceptualizing, observing, and influencing social–ecological transitions. *Ecology and Society*, 14(2).

Flannery, K., & Marcus J 2012. *The Creation of Inequality: How Our Prehistoric Ancestors Set the Stage for Monarchy, Slavery, and Empire*. Cambridge, MA: Harvard University Press.

Foley, J.A., DeFries, R., Asner, G. P., Barford, C., Bonan, G., Carpenter, S. R., Chapin, F. S., Coe, M. T., Daily, G. C., Gibbs, H. K., et al. 2005. Global consequences of land use. *Science*, 309, 570–574.

Forsyth, T. 2008. Political ecology and the epistemology of social justice. *Geoforum*, 39, 756–764

Foucault, M. 1991. La gubernamentalidad. In R. Castel, J. Donzelot, M. Foucault, J. P. de Gaudamar, Cl. Grignon, & F. Muel (eds.). *Espacios de poder*. Barcelona: Ediciones de La Piqueta.

Francis, C.A., Lieblein, G., Gliessman, S., Breland, T.A., Creamer, N., Harwood, R., Salomonsson, L., Heleniu, J., Rickel, D., Salvador, R., et al. 2003. Agroecology: the ecology of food systems. *Journal of Sustainable Agriculture*, 22(3), 99–118.

Fréguin-Gresh, S. 2017. Agroecología y Agricultura Orgánica en Nicaragua. Génesis, institucionalización y desafíos, en PP-AL (Red Políticas Públicas en América Latina y el Caribe). *Políticas Públicas a favor de la Agroecología en América Latina y el Caribe*. Brasilia: PP-AL, pp. 174–196.

Friedmann, H. 1987. The family farm and the international food regimes. In T. Shanin (ed.). *Peasants and Peasant Societies*, 2nd ed. Oxford: Basil Blackwell, pp. 247–58.

Friedmann, H. 1993. The political economy of food: a global crisis. *New Left Review*, 197, 29–57, 30–31.

Friedmann, H. 2005. From colonialism to green capitalism: social movements and emergence of food regimes. In F.H. Buttel, & P. McMichael (eds.). *New Directions in the Sociology of Global Development, Research in Rural Sociology and Development*. Amsterdam: Elsevier, pp. 227–264.

Friedmann, H. 2007. Scaling up: bringing public institutions and food service corporations into the project for a local, sustainable food system in Ontario. *Agriculture and Human Values*, 24, 389–398.

Friedmann, H. 2016. Food Regime Analysis and Agrarian Questions. Widening the Conversation. Available at: www.iss.nl/fileadmin/ASSETS/iss/Research_and_projects/Research_networks/ICAS/57-ICAS_CP_Friedman.pdf

Friedmann, H., & McMichael, P. 1989. Agriculture and the state sytem: the rise and decline of national agricultures, 1870 to the present. *Sociologia Ruralis*, 292, 93–117.

Galvão Freire, A. 2018. Women in Brazil build autonomy with agroecology. *Farming Matters*, 341. Available at: https://farmingmatters.org/farming-matters-341-1/women-in-brazil-build-autonomy-with-agroecology/

Garbach, K., Milder, J.C., DeClerck, F.A., de Wit, M.A., Driscoll, L., & Gemmill-Herren, B. 2016. Examining multi-functionality for crop yield and ecosystem services in five systems of agroecological intensification. *International Journal of Agricultural Sustainability*, 15, 11–28.

García Trujillo, R., Tobar, E., & Gómez, F. 2009. Alimentos ecológicos para conusmo social en Andalucía. In M. González de Molina (ed.). *El desarrollo de la agricultura ecológica en Andalucía 2004–2007. Crónica de una experiencia agroecológica*. Barcelona: Icaria, pp. 195–212.

Garnett, T., Appleby, M.C., Balmford, A., Bateman, I.J., Benton, T.G., Bloomer, P., Burlingame, B., Dawkins, M., Dolan, L., Fraser, D., et al. 2013. Sustainable intensification in agriculture: premises and policies. *Science*, 341, 33–34.

Garnett, T., & Godfray, C. 2012. Sustainable intensification in agriculture. navigating a course through competing food system priorities; Food Climate Research Network and the Oxford Martin Programme on the Future of Food, University of Oxford: Oxford, UK, 2012. Available online: www.fcrn.org.uk/sites/default/files/SI_report_final.pdf (accessed January 12, 2014).

Garrabou, R., & González de Molina, M. (eds.). 2010. *La reposición de la fertilidad en los sistemas agrarios tradicionales*. Barcelona: Icaria.

Garrido, F. (Comp.). 1993. *Introducción a la Ecología Política*. Granada: Editorial Comares.

Garrido, F. 2009. El decrecimiento y la soberanía popular como procedimiento. http://congresos.um.es/sefp/sefp2009/paper/viewFile/3631/3611

Garrido, F. 2014. Topofilia, paisaje y sostenibilidad del territorio. *Enrahonar. Quaderns de Filosofia*, 53, 63–75.

Garrido, F. 2015. Crisis, democracia y decrecimiento. In G.M. Bester (ed.). *Direito e ambiente uma democracia sustentavel. Diálogos multidisciplinares entre Portugal y Brasil*. Curitiva: Instituto Memorias.

Garrido, F., González de Molina, M., Serrano, J.L., & Solana, J.L. (eds.). 2007. *El paradigma ecológico en las ciencias sociales*. Barcelona: Icaria.

Garrido Peña, F. 1996. *La ecología política como política del tiempo*. Granada: Comares.

Garrido Peña, F. 2012. Ecología política y Agroecología: marcos cognitivos y diseño institucional. *Agroecología*, 6, 21–28.

Garvey, M. 2016. Novel Ecosystems, Familiar Injustices: The Promise of Justice-Oriented Ecological Restoration. Darkmatter Journal: In the Ruins of Imperial Culture 13, 1–16.

Geels, F.W. 2002. Technological transitions as evolutionary reconfiguration processes: a multi-level perspective and a case-study. *Research Policy*, 31, 1257–1274.

George, S. 2010. Converging crises: reality, fear and hope. *Globalizations*, 7, 17–22. doi:10.1080/14747731003593018.

Georgescu-Roegen, N. 1971. *The Entropy Law and the Economic Process*. Cambridge, MA: Harvard University Press.

Gezon, L., & Paulson, S. 2005. Place, power, difference: multiscale research at the dawn of the twenty first century. In S. Paulson, & L. Gezon (eds.). *Political Ecology Across Spaces, Scales and Social Groups* (pp 1–16). New Brunswick, NJ: Rutgers University Press.

Giampietro, M., Allen, T.F.H., & Mayumi, K. 2006. The epistemological predicament associated with purposive quantitative analysis. *Ecological Complexity*, 3, 307–327.

Giampietro, M., Aspinall, R.J., Ramos-Martin, J., & Bukkens, S.G.F. (eds.). 2014. *Resource Accounting for Sustainability Assessment: The Nexus between Energy, Food, Water and Land use*. Abingdon: Routledge.

Giampietro, M., Mayumi, K., & Ramos-Martin, J. 2008. Multi-Scale Integrated Analysis of Societal and Ecosystem Metabolism (MUSIASEM). An Outline of Rationale and Theory. Document de Treball. Departament d'Economia Aplicada.

Giddens, A. 1987. The Nation State and Violence. *A Contemporary Critique of Historical Materialism*, Vol. 2. Cambridge: Polity Press.

Giddens, A. 2000. *Mundo na era da globalização*. Lisbon: Presença.

Gierlinger, S., & Krausmann, F. 2012. The physical economy of United States of America. *Journal of Industrial Ecology*, 16, 365–377.

Giljum, S., & Eisenmenger, N. 2003. *North–South Trade and the Distribution of Environmental Goods and Burdens*. SERI Studies 2.

Gintis, H. 2006. *Moral Sentiments and Material Interests: The Foundations of Cooperation in Economic Life*. Cambridge, MA: MIT Press.

Gintis, H. 2009. *The Bounds of Reason: Game Theory and the Unification of the Behavioral Sciences*. Englewood Cliffs: Princeton University Press.

Giraldo, O.F. 2018. *Ecología política de la agricultura. Agroecología y posdesarrollo*. San Cristóbal de Las Casas: Del Colegio de la Frontera Sur.

Giraldo, O.F., & Rosset, P. 2016. La Agroecología en una encrucijada: entre la institucionalidad y los movimientos sociales. *Guaju*, 2 14. doi:10.5380/guaju.v2i1.48521.

Glansdorff, P., & Prigogine I. 1971. *Thermodynamic Theory of Structure, Stability and Fluctuations*. New York: Wiley Interscience.

GLASOD (The Global Assessment of Human Induced Soil Degradation). 1991. UNEP.

Gliessman, S.R. 1998. *Agroecology: Ecological Processes in Sustainable Agriculture*. Chelsea, MI: Ann Arbor Press.

Gliessman, S.R. 2011. Agroecology and food system change. *Journal of Sustainable Agriculture*, 35, 345–349.

Gliessman, S.R. 2013. Agroecología: plantando las raíces de la resistencia. *Agroecología*, 8, 19–26.

Gliessman, S. 2014. Agroecology and social transformation. *Agroecology and Sustainable Food Systems*, 38, 1125–1126. doi:10.1080/21683565.2014.951904.

Gliessman, S.R. (ed.). 2015. *Agroecology: The Ecology of Sustainable Food Systems*, 3rd ed. New York: CRC Press.

Gliessman, S. 2018. Scaling-out and scaling-up agroecology. *Agroecology and Sustainable Food Systems*, 42, 841–842.

Gliessman, S.R., Rosado-May, F.J., Guadarrama-Zugasti, C., Jedlicka, J., Cohn, A., Méndez, V.E., Cohen, R., Trujillo, L., Bacon, C.M., & Jaffe, R. 2007. Agroecología: promoviendo una transición hacia la sostenibilidad. *Ecosistemas*, 16, 13–23.

Gonsalves, J.F. 2001. Going to scale: what we have garnered from recent workshops. *LEISA Magazine*. Available from: www.agriculturesnetwork.org/library/63894

González de Molina, M. 2001. Introducción. In M. González de Molina (ed.). *La Historia de Andalucía a debate (I. Campesinos y Jornaleros)*. Barcelona: Editorial Anthropos.

González de Molina, M. (ed.). 2009. *El desarrollo de la agricultura ecológica en Andalucía 2004–2007. Crónica de una experiencia agroecológica*. Barcelona: Icaria.

González de Molina, M. 2013. Agroecology and politics. How to get sustainability? About the necessity for a political agroecology. *Agroecology and Sustainable Food Systems*, 37, 45–59.

González de Molina, M., & Guzmán-Casado, G. 2006. *Tras los Pasos de la Insustentabilidad. Agricultura y medio ambiente en perspectiva histórica*. Barcelona: Icaria.

González de Molina, M., & Guzmán Casado, G.I. 2017. Agroecology and ecological intensification. A discussion from a metabolic point of view. *Sustainability*, 9, 86.

González de Molina, M., Guzmán Casado, G., García Ruiz, R., Soto Fernández, D., Herrera González de Molina, A., & Infante Amate, J. 2010. Claves del crecimiento agrario: la reposición de la fertilidad en la agricultura andaluza de los siglos XVIII y XIX. In

R. Garrabou, & M. González de Molina (eds.). *La reposición de la fertilidad en los sistemas agrarios tradicionales*. Barcelona: Editorial Icaria.

González de Molina, M., & Sevilla Guzmán, E. 1993a. Ecología, campesinado e historia: para una reinterpretación del desarrollo del capitalismo en la agricultura. In E. Sevilla Guzmán, & M. González de Molina (eds.). *Ecología, campesinado e historia*. Madrid: Ediciones de la Piqueta, pp. 23–130.

González de Molina, M., & Sevilla Guzmán, E. 1993b. Una propuesta de diálogo entre socialismo y ecología: el neopopulismo ecológico. *Ecología Política*, 3, 121–135.

González de Molina, M. & Sevilla Guzmán, E. 2001. Perspectivas socio-ambientales de la historia del movimiento campesino andaluz. in M. González de Molina (ed.). *La Historia de Andalucía a debate. I. Campesinos y jornaleros*. Barcelona: Editorial Anthropos, pp. 239–287.

González de Molina, M, Soto Fernández, D., & Garrido Peña, F. 2016. Los conflictos ambientales como conflictos sociales. Una mirada desde la ecología política y la historia. *Ecología Política*, 50, 31–38.

González de Molina, M., Soto Fernández, D., Guzmán Casado, G., Infante Amate, J., Aguilera Fernández, E., Vila Traver, J., & García Ruiz, R. 2019. *The Agrarian Metabolism of Spanish Agriculture, 1900–2008. The Mediterranean way towards industrialization*. Berlin: Springer.

González de Molina, M., Soto Fernández, D., Infante-Amate, J., Aguilera Fernández, E., Vila Traver, J, & Guzmán Casado, G. 2017. Decoupling food from land: the evolution of Spanish agriculture from 1960 to 2010. *Sustainability*, 9, 2348.

González de Molina, M., & Toledo, V. 2011. *Metabolismos, naturaleza e historia. Una teoría de las transformaciones socioecológicas*. Barcelona: Icaria.

González de Molina, M. & Toledo, V. 2014. *The Social Metabolism: A Socioecological Theory of Historical Change*. Berlin: Springer.

Goodman, D. 2009. Place and space in alternative food networks: connecting production and consumption. Working paper #21; Environment, Politics and Development Working Paper Series. Department of Geography, King's College London.

Goodman, N. 2013. *Maneras de Hacer mundos*. Madrid: Editorial Antonio Machado S.A.

Goody, J. 1986. *La evolución de la familia y del matrimonio en Europa*. Barcelona: Herder Editorial.

Gould, F., Brown, Z.S., & Kuzma, J. 2018. Wicked evolution: can we address the sociobiological dilemma of pesticide resistance? *Science*, 360(6390), 728–732.

Gould, S.J., & Vrba, E.S. 1982. Exaptation – a missing term in the science of form. *Paleobiology*, 8, 4–15.

Graeub, B.E., Jahi Chappell, M.J., Wittman, H., Ledermann, S., Bezner Kerr, B., & Gemmill-Herren, B. 2016. The state of family farms in the world. *World Development*, 87, 1–15.

GRAIN. 2018. Emisiones imposibles. Cómo están calentando el planeta las grandes empresas de carne y lácteos. GRAIN and Institute for Agriculture and Trade Policy (www.grain.org/es) (accessed November 28, 2018).

Greenpeace International. 2018. *Less is More: Reducing Meat and Dairy for a Healthier Life and Planet*. London: Greenpeace.

Guaman, V. 2015. *Democracia deliberativa en comunidades indígenas bajo los postulados de Habermas*. Academia. Available at www.academia.edu/29580373/Democracia_deliberativa_en_comunidades_ind%C3%ADgenas_bajo_los_postulados_de_Habermas.

Guthman, J. 2004. The trouble with 'organic lite' in California: a rejoinder to the 'conventionalisation' debate. *Sociologia Ruralis*, 44, 301–316. doi:10.1111/j.1467-9523.2004. 00277.x.

Guzmán, G.I., González de Molina, M., & Sevilla, E. (eds.). 2000. *Introducción a la Agroecología como desarrollo rural sostenible*. Madrid: Mundi-Prensa.

Guzmán Casado, G.I., & González de Molina, M. 2009. Preindustrial agriculture versus organic agriculture. The land cost of sustainability. *Land Use Policy*, 26, 502–510.

Guzmán Casado, G., & González de Molina, M. 2017. *Energy in Agroecoystems. A Tool for Assessing Sustainability*. Boca Raton: CRC Press.

Guzmán Casado, G.I., González de Molina, M., & Alonso Mielgo, A. 2011. The land cost of agrarian sustainability. An assessment. *Land Use Policy*, 28, 825–835.

Habermas, J. 2010. *Faticidad y validez*. Valencia: Trotta.

Hacyan, S. 2004. *Física y metafísica del espacio y el tiempo: la filosofía en el laboratorio*. Mexico: FCE.

Hall, C.A.S. 2011. Introduction to special issue on new studies in EROI (energy return on investment). *Sustainability* 3, 1773–1777.

Hall, C.A.S., Balogh, S., & Murphy, D.J. 2009. What is the minimum EROI that a sustainable society must have? *Energies*, 2, 25–47.

Hardin, G. 1968. The Tragedy of the Commons. *Science*, 162(3859), 1243–1248.

Hardisty, D.J., & Weber, E.U. 2009. Discounting future green: money vs the environment. *Journal of Experimental Psychology: General*, 138, 329–340.

Hardt, M., & Negri, A. 2000. *Empire*. Cambridge, MA: Harvard University Press.

Harich, W. 1978. *¿Comunismo sin crecimiento? Babeuf y el club de Roma*. Barcelona: Materiales.

Harper, A., Shattuck, A., & Holt-Giménez, E. 2009. *Food Policy Councils: Lessons Learned, Food First*. Institute for Food and Development Policy.

Harris, J. 1982. *Rural Development: Theories of Peasant Economy and Agrarian Change*. London: Hutchinson University Library.

Hauser, O.P., Rand, D.G., Peysakhovich, A., & Nowak, M.A. 2014. Cooperating with the future. *Nature*, 511(7508), 220–223.

Hayek, F. 1944. *The Road to Serfdom*. Cambridge: Routledge Press, Sciences.

Hebinck, P., Ploeg, J.D. van der, & Schneider, S. (orgs.) 2015. *Rural Development and the Construction of New Markets*. New York: Routledge.

Heinrich Böll Foundation, Rosa Luxemburg Foundation, Friends of the Earth Europe. 2017. *Agrifood Atlas. Facts and figures about the corporations that control what we eat*. Berlin: Heinrich Böll Foundation.

Hendrich, J. 2017. *The Secret of Our Success How Culture Is Driving Human Evolution, Domesticating Our Species, and Making Us Smarter*. Princeton: Princeton University Press.

Hibbard, K.A., Crutzen, P.J., Lambin, E.F., Liverman, D.M., Mantua, N.J., McNeill, J.R., Messerli, B., & Steffen, W. 2007. Decadal-scale interactions of humans and the environment. In R. Costanza, et al. (eds.) *Sustainability or Collapse? An Integrated History and Future of People on Earth*. Cambridge, MA: The MIT Press, pp. 341–377.

Hirst, E., 1974. Food-related energy requirements. *Science*, 184(4133), 134–138.

HLPE. 2012. *Climate Change and Food Security; A Report by the High Level Panel of Experts on Food Security and Nutrition of the Committee on World Food Security*. Rome: FAO.

Ho, M.-W. 2013. Circular thermodynamics of organisms and sustainable systems. *Systems*, 1, 30–49.

Ho, M.-W., & Ulanowicz, R. 2005. Sustainable systems as organisms? *BioSystems*, 82, 39–51.

Hobbes, T. 1984. *Leviatán*. Mexico D.F.: Fondo de Cultura Económica.

Hobsbawn, E. 1994. *Age of Extremes: The Short Twentieth Century, 1914–1991*. London: Michael Joseph, pp. 288–289.

Holloway, J. 2011. *Agrietar el capitalismo: el hacer contra el trabajo*. Buenos Aires: Herramienta.

Holt-Giménez, E. 2001. Scaling up sustainable agriculture. Lessons from the Campesino a Campesino movement. *LEISA* magazine, October.

Holt-Giménez, E. 2006. *Campesino a Campesino: voices from Latin America's farmer to farmer movement for sustainable agriculture.* Oakland, CA: Food First Books.

Holt-Giménez, E., & Altieri, M.A. 2013. Agroecology, food sovereignty, and the new Green Revolution. *Agroecology and Sustainable Food Systems*, 371, 90–102.

Hoppe, R. 2010. *The Governance of Problems: Puzzling, Powering, Participation.* Bristol: The Policy Press.

Horlings, L.G., & Marsden, T.K. 2011. Towards the real Green Revolution? Exploring the conceptual dimensions of a new ecological modernisation of agriculture that could 'feed the world'. *Global Environmental Change*,21, 441–452. doi:10.1016/j.gloenvcha.2011.01.004.

Hornborg, A. 2007. Footprints in the cotton fields: the industrial revolutionas time-space appropiation and environmental load displacement. In A. Hornborg, J.R. McNeill & J. Martínez-Alier (eds.). *Rethinking Environmental History. World-system History and Global Environmental Change.* Lanhan: Altamira Press, pp. 259–272.

Hornborg, A. 2011. *A Lucid Assessment of Uneven Development as a Result of the Unequal Exchange of Time and Space.* Lund University Centre of Excellence for Integration of Social and Natural Dimensions of Sustainability (LUCID Assessment No. 1, November 2011).

Hufty, M. 2008. Una propuesta para concretizar el concepto de gobernanza: el Marco Analítico de la Gobernanza. In H.t Mazurek (ed.). *Gobernabilidad y gobernanza en los territorios de América Latina.* La Paz: IFEA-IRD.

IAASTD. 2009. Agriculture at a Crossroads. Global Report. International Assessment of Agricultural Knowledge, Science and Technology for Development.

INE (Instituto Nacional de Estadística). 2009. Censo agrario de 2009. www.ine.es (accessed October 4, 2016).

Infante, J., & González de Molina, M. 2010. Agroecología y decrecimiento. Una alternativa a la configuración actual del sistema agroalimentario español. *Revista de Economía Crítica*, 10, 113–137.

Infante-Amate, J., & González de Molina, M. 2013. 'Sustainable degrowth' in agriculture and food: an agro-ecological perspective on Spain's agri-food system (year 2000). *Journal of Cleaner Production*, 38, 27–35.

Infante-Amate, J., Aguilera, E., & González de Molina, M. 2014a. La gran transformación del sector agroalimentario español. Un análisis desde la perspectiva energética 1960–2010. *Sociedad Española de Historia Agraria.* DT-SEHA 1403. https://ideas.repec.org/p/seh/wpaper/1403.html (accessed April 2, 2015).

Infante-Amate, J., Aguilera, E., & González de Molina, M. 2018a. Energy transition in agri-food systems. Structural change, drivers and policy implications (Spain, 1960–2010). *Energy Policy*, 122, 570–579.

Infante-Amate, J., Aguilera, E., Palmeri, F., Guzmán, G.I., Soto, D., García-Ruiz, R., & González de Molina, M. 2018b. Land embodied in Spain's biomass trade and consumption 1900–2008. Historical changes, drivers and impacts. *Land Use Policy*, 78, 493–502.

Infante-Amate, J., & Iriarte Goñi, I. 2017. Las bioenergías en España. Una serie de producción, consumo y stocks entre 1860 y 2010.

Infante-Amate, J., Soto, D., Iriarte Goñi, I., Aguilera, E., Cid, A., Guzmán, G.I., García-Ruiz, R., & González de Molina, M. 2014b. La producción de leña en España y sus implicaciones en la transición energética. Una serie a escala provincial 1900–2000. DT-AEHE Nº1416. http://econpapers.repec.org/paper/ahedtaehe/1416.htm (accessed April 2, 2015

Infante-Amate, J., Soto Fernandez, D., Aguilera, E., García-Ruiz, R., Guzmán, G., Cid, A., & González de Molina, M. 2015. The Spanish transition to industrial metabolism. Long-term material flow analysis 1860–2010. *Journal of Industrial Ecology*, 19, 866–876. doi: DOI: 10.1111/jiec.12261.

Instituto Brasileiro de Geografia e Estatística, IBGE. 2018. *Censo Agropecuário*. Brasil, IBGE.

International Assessment of Agricultural Knowledge, Science and Technology for Development. 2009. *Synthesis Report: A Synthesis of the Global and Sub-global IAASTD Reports*. Washington, DC: IAASTD.

IPES-Food (International Panel of Experts on Sustainable Food Systems). 2016. From uniformity to diversity: a paradigm shift from industrial agriculture to diversified agroecological systems. www.ipes-food.org/agroecology (accessed November 30, 2016).

IPES-Food (International Panel of Experts on Sustainable Food Systems). 2017. Too big to feed; exploring the impacts of mega-mergers, consolidation and concentration of power in the agri-food sector. www.ipes-food.org/_img/upload/files/CS2_web.pdf (accessed January 4, 2019).

IPES-Food (International Panel of Experts on Sustainable Food Systems). 2018. Breaking away from industrial food and farming systems; seven case studies of agroecological transitions. www.ipes-food.org/_img/upload/files/CS2_web.pdf (accessed January 4, 2019).

Izumi, B.T., Wright, D.W., & Hamm, M.W. 2010. Market diversification and social benefits: motivations of farmers participating in farm to school programs. *Journal of Rural Studies*, 26, 374–382.

Jones, A., Pimbert, M., & Jiggins, J. 2011. *Virtuous Circles: Values, Systems and Sustainability*. London: IIED; IUCN; CEESP.

Jørgensen, S.E., & Fath, B.D. 2004. Application of thermodynamic principles in ecology. *Ecological Complexity*, 1, 267–280.

Jørgensen, S.E., Fath, B.D., Bastianoni, S., Marques, J.C., Müller, F., Nielsen, S.N., Tiezzi, E., & Ulanowicz, R.E. 2007. *A New Ecology: Systems Perspective*. Amsterdam: Elsevier.

Kantorowicz, E.H. 1985. *Los dos cuerpos del rey. Un estudio sobre teología política medieval*. Madrid: Alianza Editorial.

Kay, J.J., Regier, A.H., Boyle, M., & Francis, G. 1999. An ecosystem approach for sustainability: addressing the challenge for complexity. *Futures*, 31, 721–742.

Kearney, M. 1996. *Reconceptualizing the Peasantry*. Colorado: Westview Press.

Koning, N. 1994. *The Failure of Agrarian Capitalism Agrarian Politics in the UK, Germany, the Netherlands and the USA, 1846–1919*. New York: Routledge.

Krausmann, F., Erb, K.-H., Gingrich, S., Lauk, C., & Haberl, H. 2008a. Global patterns of socioeconomic biomasss flows in the year 2000: a comprehensive assessment of supply, consumption and constraints. *Ecological Economics*, 65, 471–487.

Krausmann, F., Gingrich, S., Eisenmenger, N., Erb, K.-H., Haberl, H., & Fischer-Kowalski, M. 2009. Growth in global materials use, GDP and population during the 20th century. *Ecological Economics*, 68, 2696–2705. doi: http://dx.doi.org/10.1016/j.ecolecon.2009.05.007.

Krausmann, F., Gingrich, S., & Nourbakhch-Sabet, R. 2011. The metabolic transition in Japan. *Journal of Industrial Ecology*, 15, 877–892.

Krausmann, F., Haberl, H., Schulz, N.B., Erb, K.-H., Darge, E., & Gaube, V. 2003. Land-use change and socioeconomic metabolism in Austria Part I: driving forces of land-use changes 1950–1995. *Land Use Policy*, 201, 1–20.

Krausmann, F., & Langthaler, E. 2019. Food regimes and their trade links: a socioecological perspective. *Ecological Economics*, 160, 87–95.

Krausmann, F., Schandl, H., Eisenmenger, N., Giljum, S., & Jackson, T. 2017a. Material flow accounting: measuring global material use for sustainable development. *Annual Review of Environment and Resources*, 42, 647–75.

Krausmann, F., Schandl, H.Y., & Sieferle, R.P. 2008b. Socioecological regime transition in Austria and United Kingdom. *Ecologial Economics*, 65, 187–201.

Krausmann, F., Wiedenhofer, D., Lauka, K., Haas, W., Tanikawa, H., Fishman, T., Miatto, A., Schandld, H., & Haberl, H. 2017b. Global socioeconomic material stocks rise 23-fold over the 20th century and require half of annual resource use. *Proceedings of the National Academy of Sciences USA*, 114, 1880–1885.

Krugman, P., & Obstfeld, M. 2010. *Economía internacional: teoría y política*. Mexico: Editorial Pearson Addison.

Lachman, D.A. 2013. A survey and review of approaches to study transitions. *Energy Policy*, 58, 269–276.

Laclau, E. 2005. *La razón populista*. Buenos Aires: Fondo de Cultura Económica.

Laclau, E. 2009. Laclau en debate: postmarxismo, populismo, multitud y acontecimiento (entrevistado por Ricardo Camargo). *Revista de Ciencia Política*, 29, 815–828.

Laforge, J.M.L., Anderson, C.R., & McLacham, S.M. 2017. Governments, grassroots, and the struggle for local food systems: containing, coopting, contesting and collaborating. *Agriculture and Human Values*, 34, 663–681.

Laibman, D. 1992. *Value, Technical Change and Crisis: Explorations in Marxist Economic Theory*. Abingdon: Routledge.

Lamine, C. 2015. *La fabrique sociale et politique des paradigmes de l'écologisation. HDR de sociologie, Université de Paris Ouest Nanterre la Défense. Remaniée et publiée en 2017 sous le titre La Fabrique sociale de l'écologisation de l'agriculture*. Marseille: La Discussion.

Lamine, C., Renting, H., Rossi, A., Wiskerke, J.S.C. (Han), & Brunori, G. 2012. Agri-Food systems and territorial development: innovations, new dynamics and changing governance mechanisms. In I. Darnhofer, D. Gibbon, & B. Dedieu (orgs.). *Farming Systems Research into the 21st Century: The New Dynamic*. Dordrecht: Springer, pp. 229–56. https://doi.org/10.1007/978-94-007-4503-2_11.

Lancaster, K.J. 1966. A new approach to consumer theory. *Journal of Political Economics*, 74, 132–157.

Lang, T., & Barling, D. 2012. Food security and food sustainability: reformulating the debate. *Geographical Journal*, 178, 313–326.

Lattuca, A. 2011. La agricultura urbana como política pública: el caso de la ciudad de Rosario, Argentina. *Agroecología*, 6, 97–104.

Leach, G. 1976. *Energy and Food Production*. London: IPC Science and Tecnology.

Lee, R., & Marsden, T. 2009. The globalization and re-localization of material flows: four phases of food regulation. *Journal of Law and Society*, 36, 129–144. doi:10.1111/j.1467-6478.2009.00460.x.

Leff, E. 1986. *Ecología y Capital*. Mexico: Siglo XXI Editores.

Levidow, L., Pimbert, M., & Vanloqueren, G. 2014. Agroecological research: conforming – or transforming the dominant agro-food regime? *Agroecology and Sustainable Food Systems*, 38, 1127–1155.

Levins, R. 2006. A whole-system view of agriculture, people, and the rest of nature. In A. Cohn, J. Cook, M. Fernández, R. Reider, & C. Steward (eds.). *Agroecology and the Struggle for Food Sovereignty in the Americas*. Nottingham: Russell Press, pp. 34–49.

Livi-Bacci, M. 1999. *Historia de la población europea*. Barcelona: Editorial Crítica.

Long, N. 1986. Commoditization: thesis and antithesis. In N. Long et al., The Commoditization Debate. Wageningen: Pudoc, pp. 8–23.

Long, N. & Ploeg, J.D. van der. 1994. Heterogeneity, actor and structure: towards a reconstitution of the concept of structure. In D. Booth (ed.). *Rethinking Social Development: Theory, Research, and Practice.* Harlow: Longman, pp. 62–90.

Loos, J., Abson, D.J., Chappell, M.J., Hanspach, J., Mikulcak, F., Tichit, M., & Fischer, J. 2014. Putting meaning back into "sustainable intensification". *Frontiers in Ecology and the Environment,* 12, 356–361.

Lopes, A.P., & Jomalina, E. 2011. Agroecology: exploring opportunities from women's empowerment based on experiences from Brazil. In: *Feminist Perspectives Towards Transforming Economic Power.* Toronto, Mexico City, Cape Town: Association of Women's Rights in Development.

López-García, D., Pontijas, B., González de Molina, M., Guzmán-Casado, G.I., Delgado, M. & Infante, J. 2015. *Diagnóstico para la conexión de la distribución comercial con la producción endógena andaluza en el comercio local 2015.* Junta de Andalucía: Dirección General de Comercio.

López-Moreno, I. 2014. *Labelling the origin of food products: Towards sustainable territorial development?* PhD thesis. Wageningen: Wageningen University.

Lowder, S.K., Skoet, J., & Raney, T. 2016. The number, size, and distribution of farms, smallholder farms, and family farms worldwide. *World Development,* 87, 16–29.

Luhmann, N. 1984. *Soziale Systeme: Grundriss einer allgemeine theorie.* Frankfurt: Suhrkamp.

Luhmann, N. 1986. The autopoiesis of social systems. In F. Geyer, & J.V. Zeuwen (eds.). *Sociocybernetic Paradoxes: Observation, Control and Evolution of Self-Steering Systems.* London: Sage, pp. 172–192.

Luhmann, N. 1995. *Social Systems.* Stanford: Stanford University Press.

Luhmann, N. 1998. *Complejidad y modernidad. De la unidad a la diferencia.* Editorial Trotta, España.

Luhmann, N. 2009. *¿Cómo es posible el orden social?* Barcelona: Herder.

Luhmann, N. 2010. *Organización y decisión.* Barcelona: Herder.

Lundqvist, J., de Fraiture, C., & Molden, D. 2008. Saving water: from field to fork – curbing losses and wastage in the food chain. Stockholm International Water Institute Policy Brief, Stockholm.

Luxemburgo, R. 2010. *Reforma ou revolução?* São Paulo: Expressão Popular.

Machín B., Roque, A.M., Ávila, D.R., & Rosset, P.M. 2010. *Revolución Agroecológica: El Movimiento de Campesino a Campesino de la Anap en Cuba.* Vía Campesina: Anap.

MAGRAMA (Ministerio de Agricultura y Medio Ambiente). 2007. *La alimentación en España, 2006.* Madrid: Ministerio de Agricultura, Alimentación y Medio Ambiente.

Majone, G. 2006. Agenda setting. In M. Moran (ed.). *The Oxford Handbook of Policy Public.* New York: Oxford University Press, pp. 228–250.

MAPAMA (Ministerio de Agricultura, pesca y Alimentación). 2012. *Informe del consumo de alimentación en España 2012.* Madrid: Ministerio de Agricultura, pesca Alimentación y Medioambiente.

Margalef, R. 1980. *La biosfera. Entre la termodinámica y el juego.* Barcelona: Kairós.

Margalef, R. 1993. *Teoría de los sistemas ecológicos.* Barcelona: Universitat de Barcelona.

Margalef, R. 1995. Aplicacions del caos determinsita en ecologia. In J. Flos (ed.). *Ordre i caos en ecologia.* Barcelona: Publicacions Universitat de Barcelona, pp. 171–184.

Marsden, T. 1991. Theoretical issues in the continuity of petty commodity production. In S. Whatmore, P. Lowe, & T. Marsden (eds.). *Rural Enterprise: Shifting Perspectives on Small-Scale Production.* London: David Fulton Publishers.

Marsden, T. 2003. *The Condition of Rural Sustainability.* Assen: Royal van Gorcum.

Marsden, T., Banks, J., & Bristow, G. 2000. Food supply chain approaches: exploring their role in rural development. *Sociologia Ruralis*, 40, 424–438.

Marsden, T., & Sonnino, R. 2008. Rural development and the regional state: denying multifunctional agriculture in the UK. *Journal of Rural Studies*, 24, 422–431.

Martinez, S., Hand, M., Da Pra, M., Pollack, S., Ralston, K., Smith, T., Vogel, S., Clarck, S., Lorh, L., Low, S. & Newman, C., 2010. Local food systems; concepts, impacts, and issues. Economicy Research Report 97. United States Department of Agriculture.

Martínez-Alier, J. 2007. Marxism, social metabolism, and international trade. In A. Hornborg, J.R. McNeill, & J. Martínez-Alier (eds.). *Rethinking Environmental History. World-System History and Global Environmental Change*. Lanhan: Altamira Press, pp. 221–237.

Martínez-Alier, J., Munda, G., & O'Neill, J. 1998. Weak comparability of values as a foundation for ecological economics. *Ecological Economics*, 26, 277–286.

Martínez Torres, H., Namdar-Iraní, M., & Saa Isamit, C. 2017. Las Políticas de Fomento a la Agroecología en Chile, en PP-AL (Red Políticas Públicas en América Latina y el Caribe). In *Políticas Públicas a favor de la Agroecología en América Latina y el Caribe*. Brasilia: PP-AL, pp. 70–90.

Massukado, L.M. & Balla, J.V. 2016. Panorama dos cursos e da pesquisa em agroecologia no Brasil. COMCiência, Revista Eletrônica de Jornalismo Científico. LABJOR-UNICAMPI/SBPC, 182 (October 1–10). www.comciencia.br/comciencia/?section=8&edicao=127&id=1548

Maturana, H.R., & Varela, F.J. 1980. *Autopoiesis and Cognition. The Realization of the Living*. Boston: Boston Studies in the Philosophy and History of Science.

Mauss, M. [1925] 2011. *Ensayo sobre el don. Forma y función del intercambio en las sociedades arcaicas*. Katz.

Mavrofides, T., Kameas, A., Papageorgiou, D., & Los, A. 2011. On the entropy of social systems: a revision of the concepts of entropy and energy in the social context. *Systems Research and Behavioral Science*, 28, 353–368.

Mayer, A., Schaffartzik, A., Haas, W., & Rojas-Sepúlveda, A. 2015. Patterns of global biomass trade – implications for food sovereignty and socio-environmental conflicts. EJOLT Report No. 20, 106 pp.

Mayumi, K. 1991. Temporary emancipation from land: from the industrial revolution to the present time. Ecological Economics, 4, 35–56.

Mayumi, K., & Giampietro, M 2006. The epistemological challenge of self-modifying systems: governance and sustainability in the post-normal science era. *Ecological Economics*, 57, 382–399.

Mazoyer, M., & Roudart, L. 2010. História das agriculturas no mundo; do neolítico à crise contemporânea. São Paulo: Editora UNESP.

McMichael, P. 2006. Feeding the world: agriculture, development and ecology. In L. Panitch, & C. Leys (eds.). *Socialist Register 2007*. London: Merlin Press, pp. 170–194.

McMichael, P. 2009. A food regime genealogy. *The Journal of Peasant Studies*, 36, 139–169.

McMichael, P. 2013. *Food Regimes and Agrarian Questions*. Rugby: Practical Action Publishing.

McNeill, J.R. 2000. *Something New under the Sun: An environmental history of the Twentieth Century World*. London: Penguin Books.

Méndez, V.E., Bacon, C.M., & Cohen, R. 2013. Agroecology as a transdisciplinary, participatory, and action-oriented approach. *Agroecology and Sustainable Food Systems*, 37, 3–18.

Méndez, V.E., Bacon, C.M., Cohen, R., & Gliessman, S.R. (eds.). 2016. *Agroecology: A Transdisciplinary, Participatory and Action-oriented Approach*. Boca Raton: CRC Press.

Mendras, H. 1967. *La fin des paysans – innovations et changement dans l'agriculture Francaise*. Paris : Futuribles/SEDEIS.

Meny, I., & Thoenig, J.C. 1992. *La aparición de los Problemas Públicos. Las Políticas Públicas*. España: Ariel Ciencia Política.

Mier y Terán Giménez Cacho, M., Giraldo, O.F., Aldasoro, M., Morales, H., Ferguson, B.G., Rosset, P., Khadse, A., & Campos, C. 2018. Bringing agroecology to scale: key drivers and emblematic cases. *Agroecology and Sustainable Food Systems*, 42, 637–665. DOI: 10.1080/21683565.2018.1443313

Monteiro, C.A. 2009. Nutrition and health. The issue is not food, nor nutrients, so much as processing. *Public Health Nutrition*, 12, 729–731.

Monteiro, C.A., & Cannon, G. 2012. The impact of transnational "big food" companies on the south: a view from Brazil. *PLoS Medicine*, 9(7), e1001252.

Monteiro, C.A., Levy, R.B., Claro, R.M., Ribeiro de Castro, I.R., & Cannon, G. 2010. A new classification of foods based on the extent and purpose of their processing. *Cadernos De Saude Publica*, 26, 2039–2049.

Monteiro, C.A., Moubarac, J.C., Cannon, G., Ng, S. W., & Popkin, P. 2013. Ultra-processed products are becoming dominant in the global food system. *Obesity*, 14(Suppl. 2), 21–28.

Mooney, P. 2018. Blocking the chain. Industrial food chain concentration. Big Data platforms and food sovereignty solutions. ETC Group, Canada.

Moore, B. 1966. Social origins of dictatorship and democracy. In *Lord and Peasant in the Making of the Modern World*. London: Penguin Books.

Moore, J.W. 2015. Capitalism in the web of life. In *Ecology and the Accumulation of Capital*. London: Verso.

Moors, E.H.M., Rip, A., & Wiskerke, J.S.C. 2004. The dynamics of innovation; multilevel co-evolutionary perspective. In J.S.C. Wiskerke, & J.D. van der Ploeg (eds.). *Seeds of Transition; Essays on Novelty Production, Niches and Regimes in Agriculture*. Assen: Van Gorgu.

Moran, W. 2017. Políticas a favor de la producción orgánica y agroecología en El Salvador, en PP-AL (Red Políticas Públicas en América Latina y el Caribe. 2017). In *Políticas Públicas a favor de la Agroecología en América Latina y el Caribe*. Brasilia: PP-AL, pp. 132–146.

Morgan, S.L. 2011. Social learning among organic farmers and the application of the communities of practice framework. *The Journal of Agricultural Education and Extension*, 17, 99–112.

Morgan Állman, J. 2003. *El cerebro en Evolución*. Madrid: Ariel Neurociencia.

Morin, E. 1977. *El método, I: La naturaleza de la naturaleza*. Madrid: Cátedra.

Morin, E. 2007. O método 6: Ética. Porto Alegre: Sulina.

Morin, E. 2010. Eloge de la métamorphose. *Le Monde*. www.lemonde.fr/idees/article/2010/01/09/eloge-de-la-metamorphose-par-edgar-morin_1289625_3232.html.

Muñoz, P., Giljum, S., & Roca, J. 2009. The raw material equivalents of international trade. *Journal of Industrial Ecology*, 13, 881–897.

Muradian, R., & Martínez Alier, J. 2001. Trade and environment: from Southern perspective. *Ecological Economics*, 36, 286–297.

Naredo, J.M. 2015. *La economía y evolución: Historia y perspectivas de las categorías básicas del pensamiento económico* (4th ed.). Madrid: Siglo XXI.

Nelson, G.C., Rosegrant, M.W., Koo, J., Robertson, R., Sulser, T., Zhu, T., Ringler, C., Msangi, S., Palazzo, A., Batka, M., et al. 2009. *Climate Change. Impact on Agriculture and Costs of Adaptation*. Washington, D.C.: International Food Policy Research Institute.

Nicholls, C.I., Altieri, M.A., & Vazquez, L. 2016. Agroecology: principles for the conversion and redesign of farming systems. *Journal of Ecosystem and Ecography*, S5, 010.

Niederle, P., & Almeida, L. 2013. A nova arquitetura dos mercados para produtos orgânicos: o debate da convencionalização. In P.A. Niederle, L. Almeida, & F.M. Vezzani (orgs.). *Agroecologia: práticas, mercados e políticas para uma nova agricultura.* Curitiba: Kairós, pp. 23–67.

Niederle, P.A. 2014. Os agricultores ecologistas nos mercados para alimentos orgânicos: contramovimentos e novos circuitos de comércio. *Sustentabilidade em Debate*, 5(3), 79–97.

Nieremberg, D., & Halweil, B. 2005. Cultivando seguridad alimentaria. In *La situación del mundo 2005. Redefiniendo la seguridad mundial.* Barcelona: Icaria, pp. 125–144.

Nigren, A., & Rikoon,S. 2008. Political ecology revisited: integration of politics and ecology does matter. *Society and Natural Resources*, 21, 767–782.

Northfield, T.D., & Ives, A.R. 2013. Coevolution and the effects of climate change on interacting species. *PLoS Biology*, 11(10), e1001685. doi:10.1371/journal.pbio.1001685.

Nowak, M.A. 2006. Five rules for the evolution of cooperation. *Science*, 314(5805), 1560–1563.

Oakland Institute. 2018. Agroecology case studies www.oaklandinstitute.org/agroecology-case-studies (accessed January 4, 2019).

OCDE/FAO. 2017. Carne. In *OCDE-FAO Perspectivas Agrícolas 2017–2026.* Paris: OECD Publishing.

Offe, C. 1990. Contradicciones en el Estado del Bienestar. Madrid: Alianza Editorial.

Okasha, S. 2006. *Evolution and the Levels of Selection.* Oxford: Oxford University Press.

Okishio, N. 1961. Technical changes and the tate of profit. *Kobe University Economic Review*, 7, 85–99.

Olson, M. 1971. *The Logic of Collective Action: Public Goods and the Theory of Groups.* Cambridge, MA: Harvard Economic Studies.

Oostindie, H., Rudolf, B., Brunori, G., & Ploeg, J.D. van der. 2008. The endogeneity of rural economies. In J.D. van der Ploeg & T. Marsden (eds.). *Unfolding Webs; The Dynamics of Regional Rural Development.* Assen: Van Gorcum, pp. 53–67

Orozco, A.P. 2004. Estrategias feministas de deconstrucción del objeto de estudio de la economía. *Foro Interno. Madrid*, 4, 87–117.

Ostrom, E. 1990. *Governing the Commons: The Evolution of Institutions for Collective Action.* Cambridge: Cambridge University Press.

Ostrom, E. 2001. Commons, institutional diversity of. In Encyclopedia of Biodiversity, Volume I. New York: Academic Press.

Ostrom, E. 2009. A general framework for analyzing sustainability of social-ecological systems. *Science*, 325, 419–422.

Ostrom, E. 2013. *Comprender la diversidad institucional.* Oviedo: KRK.

Ostrom, E. 2015a. *Governing the Commons: The Evolution of Institutions for Collective Action.* Canto classics. Cambridge: Cambridge University Press.

Ostrom, E. 2015b. Beyond markets and states: polycentric governance of complex economic systems. In D.H. Cole, & M.D. McGinnis (orgs.). *Elinor Ostrom and the Bloomington School of Political Economy: Polycentricity in Public Administration and Political Science.* Lanham, MD: Lexington Books.

Overton, M. 1991. The determinant of crop yield in early modern England. In B. Campbell, & M. Overton (eds.). *Land, Labour and Livestock. Historical Studies in European Agricultural Productivity.* Manchester: Manchester University Press.

Owen, W.F. 1966. The double developmental squeeze on agriculture. *The American Economic Review*, LVI, 43–67.

Pacífico, D.A. 2010. Avaliação de impacto das ações de formação/capacitação em Agroecologia realizadas pelo DATER/SAF, no período 2004 a 2009. Programa das

Nações Unidas para o Desenvolvimento Projeto PNUD/PRONAF II – BRA/06/010. Brasília-DF: PNUD.

Pahnke, A. 2015. Institutionalizing economies of opposition: explaining and evaluating the success of the MST's cooperatives and agroecological repeasantization. *The Journal of Peasant Studies*, 42, 1087–1107.

Parmentier, S. 2014. *Scaling-up Agroecological Approaches: What, Why and How?* Brussels: Oxfam-Solidarity.

Parson, T. 1976. *El sistema social*. Madrid: Revista de Occidente.

Patrouilleau, M., Martínez, L., Cittadini, E., & Cittadini, R. 2017. Políticas públicas y desarrollo de la agroecología en Argentina, en PP-AL (Red Políticas Públicas en América Latina y el Caribe. 2017. *Políticas Públicas a favor de la Agroecología en América Latina y el Caribe*. Brasilia: PP-AL, pp. 20–43.

Pattee, H.H. 1995. Evolving self-reference: matter, symbols, and semantic closure. *Commununication and Cognition in Artificial Intelligence*, 12, 9–28.

Paulson, S., Gezon, L., & Watts, M. 2003. Locating the political in political ecology: an introduction. *Human Organization*, 62, 205–217.

Perez-Cassarino, J. 2013. A construção social de mecanismos alternativos de mercados no âmbito da Rede Ecovida de agroecologia. Curitiba: UFPR. www.acervodigital.ufpr.br/handle/1884/27480?show=full.

Pérez Rivero, J.A. 2016. Puesta en valor de los subproductos obtenidos de la almazara Coop. Ntra. Sra. De las Virtudes y su potencial en el secuestro de carbono. Master's thesis. Master in Organic Farming. International University of Andalusia, Seville.

Perfecto, I., Vandermeer, J., & Wright, A. 2009. *Nature's Matrix: Linking Agriculture, Conservation and Food Sovereignty*. London: Earthscan.

Petersen, P. 2011. *Metamorfosis agroecológica: un ensayo sobre Agroecologia Política*. Baeza: UNIA.

Petersen, P. 2017. Arreglos institucionales para la intensificación agroecológica: una mirada al caso brasileño desde la Agroecología Política. PhD dissertation, Seville, September 2017.

Petersen, P. 2018. Agroecology and the restoration of organic metabolism in agrifood systems. In T. Marsden (ed.). *The Sage Book of Nature*. London: Sage, pp. 1448–1467.

Petersen P., Mussoi, E.M., & Dal Soglio, F. 2013. Institutionalization of the agroecological approaching Brazil: advances and challenges. *Agroecology and Sustainable Food Systems*, 37, 103–114.

Petersen, P., & Silveira, L.M. 2017. Agroecology, public policies and labor-driven intensification: alternative development trajectories in the Brazilian semi-arid region. *Sustainability*, 9, 535. doi:10.3390/su9040535.

Peterson, G. 2000. Political ecology and ecological resilience: an integration of human and ecological dynamics. *Ecological Economics*, 35, 323–336.

Petrini, C., Bogliotti, C., Rava, R., & Scaffidi, C. 2016. La centralidad del alimento. Documento congresual, 2012–2016. Slow Food: https://slowfood.com/filemanager/official_docs/SFCONGRESS2012_La_centralidad_del_alimento.pdf

Pfister, C. 1990. The early loss of ecological stability in an Agrarian region. In P. Brimblecombe, & C. Pfister (eds.). *The Silent Countdown. Essays in European Environmental History*. Berlin: Springer, pp. 37–55.

Piketty, T. 2014. *El capital en el siglo XXI*. Mexico: FCE.

Pimbert, M. 2015. Agroecology as an alternative vision to conventional development and climate-smart agriculture. *Development*, 58, 286–298.

Pimbert, M.P. 2018. Democratizing knowledge and ways of knowing for food sovereignty, agroecology, and biocultural diversity. In: M.P. Pimbert (ed.). *Food Sovereignty,*

Agroecology, and Biocultural Diversity. Constructing and Contesting Knowledge. London: Routledge.

Pimentel, D., & Pimentel, M. 1979. *Food, Energy and Society.* London: Edward Arnold,.

Ploeg, J.D. van der. 1990. *Labor, Markets, and Agricultural Production. Westview Special Studies in Agriculture Science and Policy.* Boulder: Westview Press.

Ploeg, J.D. van der. 1993. El proceso de trabajo agrícola y la mercantilización. In E. Sevilla Guzman, & M. González de Molina (eds.). *Ecología, campesinado e historia.* Madrid: Ediciones de la Piqueta, pp. 153–196.

Ploeg, J.D. van der. 2007. Resistance of the third kind and the construction of sustainability. www.jandouwevanderploeg.com/EN/publications/articles/resistance-of-the-third-kind/.

Ploeg, J.D. van der. 2008. *The New Peasants. Struggles for the Autonomy and Sustainability in an Era of Empire and Globalization.* London: Earthscan.

Ploeg, J.D. van der. 2010. The peasant mode of production revisited. *Rural Development: Challenges and Interlinkages.* www.jandouwevanderploeg.com/EN/publications/articles/the-peasant-mode-of-production-revisited/.

Ploeg, J.D. van der. 2012. The drivers of change: the role of peasants in the creation of an agroecological agriculture. *Agroecología,* 6, 47–54.

Ploeg, J.D. van der. 2013. *Peasants and the Art of Farming: A Chayanovian Manifesto.* Agrarian Change and Peasant Studies Series. Halifax: Fernwood Publishing.

Ploeg, J.D. van der. 2015. Newly emerging, nested markets. A theoretical introduction. In P. Hebinck, J.D. van der Ploeg, & S. Schneider (orgs.). *Rural Development and the Construction of New Markets.* Abingdon: Routledge, pp. 16–40.

Ploeg, J.D. van der. 2018a. From de-to repeasantization: the modernization of agriculture revisited. *Journal of Rural Studies,* 61, 236–243.

Ploeg, J.D. van der. 2018b. *The New Peasantries; Rural Development in Times of Globalization,* 2nd ed. New York: Routledge.

Ploeg, J.D. van der, Bouma, J., Rip, A., Rijkenberg, F.H.J., Ventura, F., & Wiskerke, J.S.C. 2004. On regimes, novelties, niches and co-production". In J.D. van der Ploeg, & J.S.C. Wiskerke (eds.). *Seeds of Transition : Essays on Novelty Production, Niches and Regimes in Agriculture.* European Perspectives on Rural Development. Assen: Van Gorcum, pp. 1–30. http://edepot.wur.nl/337404.

Polanyi, K. 2001. *The Great Transformation: The Political and Economic Origins of Our Time.* 2nd ed. Boston, MA: Beacon Press.

Polanyi, K. 2012. Formas de integração e estruturas de apoio. In *A subsistência do homem e ensaios correlatos.* Rio de Janeiro: Contraponto Editora, pp. 83–93.

Porto, S. 2016. *Agroecologia e o Programa de Aquisição de Alimentos* (PAA, Carta Maior, June 14, 2016). http://cartamaior.com.br/?/Editoria/Meio-Ambiente/A-agroecologia-e-o-Programa-de-Aquisicao-de-Alimentos-PAA/3/36284 (accessed March 10, 2019).

PP-AL (Red Políticas Públicas en América Latina y el Caribe). 2017. *Políticas Públicas a favor de la Agroecología en América Latina y el Caribe.* Brasilia: PP-AL. www.pp-al.org/es

Pretty, J., & Bharucha, Z.P. 2014. Sustainable intensification in agricultural systems. *Annals of Botany,* 114, 1571–1596.

Prigogine, I. 1947. *Etude Thermodynamique des Phenomenes Irreversibles.* Paris: Liège.

Prigogine, I. 1955. Thermodynamics of irreversible processes and fluctuations. *Temperature,* 2, 215–232.

Prigogine, I. 1962. *Non-equilibrium Statistical Mechanics.* New York: Interscience.

Prigogine, I. 1978. Time structure and fluctuations. *Science,* 201, 777–785.

Prigogine, I. 1983. ¿*Tan sólo una ilusión? Una exploración del caos al orden.* Barcelona: Tusquets.

Puleo, A.H. 2011. *Ecofemnismo: para otro mundo posible.* Barcelona: Catedra.

Ramos García, M., Guzmán, G.I., & González de Molina, M. 2018. Dynamics of organic agriculture in Andalusia: moving toward conventionalization? *Agroecology and Sustainable Food Systems,* 42, 328–359.

Ramos-Martin, J. 2003. Empiricism in ecological economics: a perspective from complex systems theory. *Ecological Economics,* 46, 387–398.

Ramos-Martin, J. 2012. Economía Biofísica. *Investigación y Ciencia,* June, 68–75.

Rawls, J. 1993. *El liberalismo político.* Barcelona: Crítica.

Raynolds, L.T. 2004. The globalization of organic agro-food networks. *World Development,* 32, 725–743.

Reardon, T., Timmer, C.P., Barrett, C.B., & Berdegué, J. 2003. The rise of supermarkets in Africa, Asia and Latin America. *American Journal of Agricultural Economics,* 85, 1140–1146.

Reed, M. 2009. For whom? The governance of organic food and farming in the UK. *Food Policy,* 34, 280–286.

Reher, D., & Camps, E. 1991. Las economías familiares dentro de un contexto histórico comparado. *Revista Española de Investigaciones Sociológicas,* 55, 65–91.

Renting, H., & Wiskerke, H. 2010. New emerging roles for public institutions and civil society in the promotion of sustainable local agro-food systems. 9th European IFSA Symposium, Vienna.

Renting, H., Schermer, M., & Rossi, A. 2012. Building food democracy: exploring civic food networks and newly emerging forms of food citizenship. *International Journal of Sociology of Agriculture and Food,* 19, 289–307.

Retamozo, M. 2017. La teoría del populismo de Ernesto Laclau: una introducción. *Estudios Políticos,* 41, 157–184.

Riechmann, J. 2006. *Biomimesis: ensayos sobre imitación de la naturaleza, ecosocialismo y autocontención.* Madrid: La Catarata.

Rigby, D., & Bown, S. 2003. Organic food and global trade: is the market delivering agricultural sustainability? Discussion Paper Series no. 0326. School of Economic Studies, University of Manchester.

Rockström, J., Steffen, W., Noone, K., Persson, A., Chapin, F.S., Lambin, E.F., Lenton, T.M., Scheffer, M., Folke, C., Schellnhuber, H.J., et al. 2009a. A safe operating space for humanity. *Nature,* 461, 472–475.

Rockström, J., Steffen, W., Noone, K., Persson, Å., Chapin, F.S., Lambin, E., Lenton, T.M., Scheffer, M., Folke, C., Schellnhuber, H.J., et al. 2009b. Planetary boundaries: exploring the safe operating space for humanity. *Ecology and Society,*14(2). doi:10.5751/ES-03180-140232.

Rosen, R. 1985. *Anticipatory Systems: Philosophical, Mathematical and Methodological Foundations.* New York: Pergamon Press.

Rosen, R. 2000. *Essays on Life Itself.* New York: Columbia University Press.

Rosset, P.M. 2013. Re-thinking agrarian reform, land and territory in La Vía Campesina. *Journal of Peasant Studies,* 40, 721–775.

Rosset, P. 2003. Food sovereignty: global rallying cry of farmer movements. Institute for Food and Development Policy Backgrounder, 9(4), Fall, 4 pp.

Rosset, P., & Altieri, M. 2017. *Agroecology: Science and Politics.* Agrarian Change and Peasant Studies Series. Winnipeg: Fernwood Publishing.

Rougoor, C.W., Zeijts, H. van, Hofreither, M.F., & Bäckman, S. 2001. Experiences with fertilizer taxes in Europe. *Journal of Environmental Planning and Management,* 44, 877–887.

Roy, P., Nei, D., Orikasa, T., Xu, Q., Okadome, H., Nakamura, N., & Shiina, T. 2009. A review of life cycle assessment (LCA) on some food products. *Journal of Food Engineering*, 90, 1–10.

Royal Society, The. 2009. *Reaping the Benefits: Science and the Sustainable Intensification of Global Agriculture*. London: The Royal Society, pp. 1–72.

Sabatier, P., & Jenkins-Smith, H. 1993. *Policy Change and Learning: An Advocacy Coalition Approach*. Boulder: Westview Press.

Sabatier, P.A., & Weible, C.M. 2014. *Theories of the Policy Process*. Boulder: Westview Press.

Sabourin, E.; Guéneau, S.; Colonna, J., Silva, L. T. da. (orgs). Construção de políticas de agroecologia (e produção orgânica) nos estados federados do Brasil. Rio de Janeiro, PP-AL/E-Papers. In press.

Sabourin, E., Niederle, P., Le Coq, J.F., Vázquez, L., & Patrouilleau, M.M 2017. Análisis comparativo en escala regional, en PP-AL (Red Políticas Públicas en América Latina y el Caribe. 2017. *Políticas Públicas a favor de la Agroecología en América Latina y el Caribe*. Brasilia: PP-AL, pp. 196–213.

SAF/MDA (Secretaria da Agricultura Familiar/Ministério do Desenvolvimento Agrário). 2004. *Programa Nacional de Apoio à Agricultura de Base Ecológica nas Unidades Familiares de Produção*. Brasília-DF: SAF/MDA.

Sahlins, M. 1960. *Evolution and Culture*. Ann Arbor: University of Michigan.

Sánchez, A.L. 1999. La crítica de la economía de mercado en Karl Polanyi: el análisis institucional como pensamiento para la acción. *Revista española de investigaciones sociológicas*, 86, 27–54.

Santa Marín, J.F., & Toro Betancur, A. 2015. Tribología: pasado, presente y futuro. *TecnoLógicas*, 18, 09–10. www.scielo.org.co/scielo.php?script=sci_arttext&pid=S0123-77992015000200001&lng=en&tlng=es

Santos, M. 1994. O retorno do território. In M. Santos, M.A. De Souza, & M.L. Silveira (orgs.). *Território: Globalização e Fragmentação*. São Paulo: Hucitec/Anpur, pp. 15–20.

Scartascini, C., Stein, E., & Tommasi, M. 2009. Political Institutions, Intertemporal Cooperation and the Quality of Policies. Working paper, Inter-American Development Bank, Research Department, No. 676.

Schaffartzik, A., Mayer, A, Gingrich, S., Eisenmenger, N., Loy, C. & Krausmann, F. 2014. The global metabolic transition: regional patterns and trends of global material flows, 1950–2010. *Global Environmental Change*, 26, 87–97.

Schandl, H., Grünbühel, C., Haberl, H., & Weisz, H. 2002. *Handbook of Physical Accounting. Measuring bio-physical dimensions of socioeconomic activities MFA – EFA – HANPP*. Vienna: Institute for Interdisciplinary Studies of Austrian Universities (IFF).

Schandl, H., Hatfield-Dodds, S., Wiedmann, T., Geschke, A., Cai, Y., West, J., Newth, D., Baynes, T., Lenzen, M., & Owen, A. 2016. Decoupling global environmental pressure and economic growth: scenarios for energy use, materials use and carbon emissions. *Journal of Cleaner Products*, 132, 45–56.

Schandl, H., & Krausmann, F. 2007. The great transformation: a socio-metabolic reading of the industrialization of United Kingdom. In M. Fisher-Kowalski, & H. Haberl (eds.). *Socioecological Transitions and Global Change. Trajectories of Social Metabolism and Land Use*. Cheltenham, Edward Elgar, pp. 83–115.

Scheffer, M., van Bavel, B., van de Leemput, I.A., & van Nes, E.H. 2017. Inequality in nature and society. *Proceedings of the National Academy of Sciences USA*, 114(50), 13154–13157.

Scheidel, A., & Sorman, A.H. 2012. Energy transitions and the global land rush: ultimate drivers and persistent consequences. *Global Environmental Change*, 22, 559–794.

Schelling, Th. 1989. *Micromotivos y macroconductas*. Mexico DF: Fondo de Cultura Económica.

Schlich, E., & Fleissner, U. 2005. The ecology of scale: assessment of regional energy turnover and comparison with global food. *International Journal of Life Cycle Assessment*, 10, 219–223.

Schmidhuber, J. 2006. The EU diet – evolution, evaluation and impacts of the CAP. FAO Documents, 2006. www.fao.org/fileadmin/templates/esa/Global_persepctives/Presentations/Montreal-JS.pdf

Schmitt, C., Niederle, P., Ávila, M., Sabourin, E., Petersen, P., Silveira, L., Assis, W., Palm, J., & Fernandes, G.B 2017. A experiência brasileira de construção de políticas públicas em favor da agroecologia, en PP-AL (Red Políticas Públicas en América Latina y el Caribe. 2017. *Políticas Públicas a favor de la Agroecología en América Latina y el Caribe*. Brasilia, PP-AL, pp. 44–69.

Schneider, S., & Escher, F.A 2011. Contribuição de Karl Polanyi para a sociologia do desenvolvimento rural. *Sociologias*, 13(27), 180–219. doi:10.1590/S1517-45222011000200008.

Schneidewind, U., Singer-Brodowski, M., Augenstein, K., & Stelzer, F. 2016. Pledge for a transformative science – a conceptual framework. Working Paper no. 191, Wuppertal Institute for Climate, Environment and Energy, Wuppertal, Germany.

Schor, J.B. 2005. Prices and quantities: unsustainable consumption and the global economy. *Ecological Economics*, 55, 309–320.

Schotter, A. 1981. *The Economics of Social Institutions*. Cambridge: Cambridge University Press.

Scott, J.C. 1976. *The Moral Economy of the Peasant: Rebellion and Subsistence in Southeast Asia*. New Haven: Yale University Press.

Scott, J.C.1998. *Seeing like a State: How Certain Schemes to Improve the Human Condition have Failed*. New Haven and London: Yale University Press.

Serrano, J.L. 2007. Pensar a la vez la ecología y el estado. In F. Garrido, M. González de Molina, J.L. Serrano, & J.L. Solana (eds.). *El paradigma ecológico en las ciencias sociales*. Barcelona: Icaria, pp. 155–199.

Sevilla Guzmán, E., & González de Molina, M. 1990. Ecosociología: Elementos teóricos para el análisis de la coevolución social y ecológica en la agricultura. *Revista Española De Investigaciones Sociológicas*, 52, 7–45.

Sevilla Guzmán, E., & González de Molina, M 2005. *Sobre a evoluçao do conceito de campesinato*. São Paulo: Editora de Expressao Popular.

Shanin, T. 1966. The peasantry as a political factor. *The Sociological Review*, 14, 5–27. doi:10.1111/j.1467-954X.1966.tb01148.x.

Shanin, T. 1979. Definiendo al campesinado: conceptualizaciones y desconceptualizaciones. Pasado y presente de un debate marxista. *Agricultura y Sociedad*, 11, 9–52.

Shanin, T. 1988. Expolary economies: a political economy of margin. *Connections*, 11(3), 18–22.

Shanin, T. 1990. *Defining Peasants. Essays Concerning Rural Societies, Explorary Economies, and Learning from them in the Contemporary World*. Oxford: Basil Blackwell.

Sherwood, S., Schut, M., & Leeuwis, C. 2012. Learning in the social wild: encounters between Farmer Field Schools and agricultural science and development in Ecuador. In H.R. Ojha, A. Hall, & R. Sulaiman (eds.). *Adaptive Collaborative Approaches in Natural Resources Governance: Rethinking Participation, Learning and Innovation*. London: Routledge, pp. 102–137.

Shils, E. 1956. *The Torment of Secrecy: The Background and Consequences of American Security Policies*. New York: Wiley.

Sieferle, P. 2001. Qué es la Historia Ecológica. In: M., González de Molina, & J. Martínez Alier (eds.). *Naturaleza Transformada. Estudios de Historia Ambiental en Espa~na*. Barcelona: Icaria.

Silliprandi, E. 2015. *Mulheres e Agroecologia; transformando o campo, as florestas e as pessoas*. Rio de Janeiro: URFJ.

Singer, P 2002. *One World: The Ethics of Globalization*. New Haven: Yale University Press.

Singh, S.J., Krausmann, F., Gingrich, S., Haberl, H., Erb, K.-H., & Lanz, P. 2012. India's biophysical economy, 1961–2008. Sustainability in a national and global context. *Ecological Economics*, 76, 60–69. doi:10.1016/j.ecolecon.2012.01.022.

Smil, V. 2001 [1999]. *Energías. Una guía ilustrada de la biosfera y la civilización*. Barcelona: Editorial Crítica.

Smith, A., & Raven, R. 2012. What is protective space? Reconsidering niches in transitions to sustainability. *Research Policy*, 41, 1025–1036.

Sosa, B.M., Jaime, A.M.R., Lozano, D.R.Á., & Rosset, P.M. (n.d.). *Revolución Agroecológica: el Movimiento de Campesino a Campesino de la ANAP en Cuba*. La Habana: ANAP.

Steffen, W., Broadgate, W., Deutsch, L., Gaffney, O., & Ludwig, C. 2015a. The trajectory of the Anthropocene: the great acceleration. *Anthropological Review*, 2, 81–98.

Steffen, W., Richardson, K., Rockström, J., Cornell, S.E., Fetzer, I., Bennett, E.M., Biggs, R., Carpenter, S.R., de Vries, W., de Wit, C.A., et al. 2015b. Planetary boundaries: guiding human development on a changing planet. *Science*, 347(6223). doi:10.1126/science.1259855

Steinhart, J.S., & Steinhart, C.E. 1974. Energy use in the US food system. *Science*, 184(4134), 307–316.

Stoker, G. 1998. Governance as theory: five propositions. *International Social Science Journal*, 50(155), 17–28.

Subirats, J., Knoepfel, P., Larrue, C., & Varone, F. 2012. *Análisis y gestión de políticas públicas*. Barcelona: Ariel.

Swannack, T.M., & Grant, W.E. 2008. Systems ecology. *Encyclopedia of Ecology*, 3477–3481.

Swanson, G.A., Bailey, K.D., & Miller, J.G. 1997. Entropy, social entropy and money: a living systems theory perspective. *Systems Research and Behavioral Science*, 141, 45–65.

Tainter, J.A. 1988. *The Collapse of Complex Societies*. Cambridge: Cambridge University Press.

Tainter, J. 2007. Scale and dependency in World Systems: local societies in convergent evolution. In A. Hornborg, J.R. McNeill, & and J. Martínez-Alier (eds.). *Rethinking Environmental History*. Lanhan: Altamira Press, pp. 361–378.

Tapia, J.A., & Astarita, R. 2011. *La Gran Recesión y el capitalismo del siglo XXI*. Madrid: Libros La Catarata.

Tello, E., Garrabou, R., Cussó, X., & Olarieta, J.R. 2008. Una interpretación de los cambios de uso del suelo desde el punto de vista del metabolismo social agrario. La comarca catalana del Vallès, 1853–2004. *Revista Iberoamericana de Economía Ecológica*, 7, 97–115

Tello, E., Garrabou, R., Cussó, X., & Olarieta, J.R. 2010. Sobre la sostenibilidad de los sistemas agrarios. Balances de nutrientes y sistemas de fertilización en la agricultura catalana a mediados del siglo XIX. In R. Garrabou, & M. González de Molina (eds.). *La reposición de la fertilidad en los sistemas agrarios tradicionales*. Barcelona: Icaria.

Tello, E., Galán, E., Sacristán, V., Moreno, D., Cunfer, G., Guzmán, G., González de Molina, M., Krausmann, F., & Gingrich, S. 2014. A proposal for a workable analysis of Energy Return on Investment (EROI). in agroecosystems. Social Ecology Working Paper 14, Vienna.

Thirsk, J. 1997. *Alternative Agriculture: A History – From the Black Death to the Present Day*. New York: Oxford University. Press.

Tilman, D., Balzer, C., Hill, J., & Befort, B.L. 2011. Global food demand and the sustainable intensification of agriculture. *Proceedings of the National Academy of Sciences USA*, 108, 20260–20264.

Tilman, D., & Clark, M. 2014. Global diets link environmental sustainability and human health. *Nature*, 515, 518–522.

Tilman, D., Fargione, J., Wolff, B., D'Antonio, C., Dobson, A., Howarth, R., Schindler, D., Schlesinger, W.H., Simberloff, D., & Swackhamer, D. 2001. Forecasting agriculturally driven global environmental change. *Science*, 292, 281–284.

Tittonell, P., Klerkx, L., Baudron, F., Félix, G.F., Ruggia, A., Apeldoorn, D., Dogliotti, S., Mapfumo, P., & Rossing, W.A.H. 2016. Ecological intensification: local innovation to address global challenges. *Sustainable Agriculture Reviews*, 19, 1–34. doi:10.1007/978-3-319-26777-7_1.

Tirole, J. 2016. *Économie du bien commun*. Paris: Presses universitaires de France.

Toledo, V.M. 1990. «The ecological rationality of peasant production», M. Altieri y S. Hecht (eds. *Agroecology and Small Farm Production*, CRC Press, USA, 53–60.

Toledo, V.M. 1993. La racionalidad ecológica de la producción campesina. In E. Sevilla, & M. González de Molina (eds.). *Ecología, campesinado e Historia*. Madrid: La Piqueta, pp. 197–218.

Toledo, V.M. 1994. La Apropiación Campesina de la Naturaleza: una aproximación etno-ecológica. Tesis de Doctor en Ciencias, Facultad de Ciencias, Universidad Nacional Autónoma de México.

Toledo, V.M. 1995. Campesinidad, agroindustrialidad, sostenibilidad: los fundamentos ecológicos e históricos del desarrollo rural. In *Cuadernos de trabajo del grupo interamericano para el desarrollo sostenible de la agricultura y los recursos naturales*. No. 3, 29 pp.

Toledo, V.M. 1999. Las "disciplinas híbridas": 18 enfoques interdisciplinarios sobre naturaleza y sociedad. *Persona y Sociedad*, XIII, 21–26.

Toledo, V.M. 2012a. Los grandes problemas ecológicos, in A. Bartra (ed.). *Los Grandes Problemas Nacionales*. Barcelona: Editorial Itaca, pp. 29–34.

Toledo, V.M. 2012b. La Agroecología en Latinoamérica: tres revoluciones, una misma transformación. *Agroecología*, 6, 37–46.

Toledo, V.M. 2015. COP 21: entre el mono pensante y el mono demente. *La Jornada*. www.jornada.unam.mx/2015/12/08/opinion/020a2pol.

Toledo, V.M., & Barrera-Bassols, N. 2008. *La Memoria Biocultural. La importancia agroecológica de las sabidurías tradicionales*. Barcelona: Editorial Icaria.

Toledo, V.M., & Barrera Bassols, N. 2017. Political agroecology in Mexico: a path toward sustainability. *Sustainability*, 92, 268.

Toledo, V.M., Carabias, J., Mapes, C., & Toledo, C. 1985. *Ecología y autosuficiencia alimentaria*. Mexico: Siglo XXI.

Torres-Melo, J., & Santander, J. 2013. *Introducción a las políticas públicas Conceptos y herramientas desde la relación entre Estado y ciudadanía*. Bogota: IEMP ediciones.

Tyrtania, L. 2008. La indeterminación entrópica Notas sobre disipación de energía, evolución y complejidad. *Desacatos*, 28, 41–68.

Tyrtania, L. 2009. *Evolución y sociedad: termodinámica de la supervivencia para una sociedad a escala humana*. Mexico: Universidad Autónoma Metropolitana.

UK Government Office for Science. Foresight 2011. The Future of Global Food and Farming; Final Project Report. London: Government Office for Science London.

Ulanowicz, R.E. 1983. Identifying the structure of cycling in ecosystems. *Mathematical Biosciences*, 65, 210–237.

Ulanowicz, R.E. 2004. On the nature of ecodynamics. *Ecological Complexity*, 1, 341–354.

UN. 2015a. Paris Agreement: United Nations Framework Convention on Climate Change. http://unfccc.int/files/essential_background/convention/application/pdf/english_paris_agrement.pdf.

UN. 2015b. Transforming our world: the 2030 agenda for sustainable development. UN. https://sustainabledevelopment.un.org/content/documents/21252030%20Agenda%20for%20Sustainable%20Development%20web.pdf.

UN. 2016. New UN Decade aims to eradicate hunger, prevent malnutrition. UN News Centre. May 4, 2016. www.un.org/apps/news/story.asp?NewsId=53605#.WaCudsaQyJC.

UNEP (United Nations Environment Programme). 1991. *World Map of Status of Human-induced Soil Degradation. An Explanatoy Note.* Wageningen: UNEP/ISRC.

UNEP. 1994. *The Pollution of Lakes and Reservoirs.* Kenya: UNEP.

UNEP. 2010. *Assessing the Environmental Impacts of Consumption and Production. Priority Products and Materials.* Paris: UNEP.

UNEP. 2011. *Decoupling Natural Resource Use and Environmental Impacts from Economic Growth. A Report of the Working Group on Decoupling to the International Resource Panel.* Lausanne: UNEP.

UNEP. 2012. The end to cheap oil: a threat to food security and an incentive to reduce fossil fuels in agriculture. UNEP Sioux Falls. http://na.unep.net/geas/getUNEPPageWithArticleIDScript.php?article_id=81 (accessed August 30, 2015).

UNEP. 2016. Global Material Flows and Resource Productivity. Assessment Report for the UNEP International Resource Panel.

United Nations Conference on Trade and Development. 2013. *Trade and Environment Review 2013: Wake Up before it is Too Late: Make Agriculture Truly Sustainable Now for Food Security in a Changing Climate.* Geneva: Author.

Valle Rivera, M. del C. del, & Martínez, J.M. T. (orgs.). 2017. *Gobernanza territorial y Sistemas Agroalimentarios Localizados en la nueva ruralidad.* Mexico: Red de Sistemas Agroalimentarios Localizados (Red Sial-México).

Vázquez, L., Marzin, J., & González, N. 2017. Políticas públicas y transición hacia la agricultura sostenible sobre bases agroecológicas en Cuba, en PP-AL (Red Políticas Públicas en América Latina y el Caribe. 2017. *Políticas Públicas a favor de la Agroecología en América Latina y el Caribe.* Brasilia: PP-AL, pp. 108–131.

Ventura, F., Brunori, G., Milone, P., & Berti, G. 2008. The rural web: a synthesis. In J.D. van der Ploeg, & T. Marsden (eds.). Unfolding Webs; The Dynamics of Regional Rural Development. Assen: Van Gorcun.

Vilhena, D.A., & Antonelli, A. 2015. A network approach for identifying and delimiting biogeographical regions. *Nature Communications*, 6, 6848.

Walker, P.A. 2007. Political ecology: where is the politics? *Progress in Human Geography*, 31, 363–369.

Wallerstein, I. 1974. *The Modern World System. Capitalist Agriculture and the Origins of the European World-economy in the Sixteenth Century.* London: Academic Press.

Wallerstein, I. 2005. Análisis de sistemas-mundo, una introducción. Mexico: Siglo XXI Editores.

Warde, P. 2009. The environmental history of pre-industrial agriculture in Europe. In S. Sörlin, & P. Warde (eds.). *Nature's End: History and Environment.* New York: Palgrave Macmillan Press.

Weis, T. 2013. *The Ecological Hoofprint: The Global Burden of Industrial Livestock.* London: Zed Books.

Wezel, A., Bellon, S., Doré, T., Francis, C., Vallod, D., & David, C. 2009. Agroecology as a science, a movement and a practice. A review. *Agronomy for Sustainable Development*, 29, 503–515.

Wironen, M.B., Bartlett, R., & Erickson, J.D. 2019. Deliberation and the promise of a deeply democratic sustainability transition. *Sustainability*, 11, 1023.

Wiskerke, J.S.C. 2009. On places lost and places regained: reflections on the alternative food geography and sustainable regional development. *International Planning Studies*, 14, 369–387. doi:10.1080/13563471003642803.

Witzke, H., & Noleppa, S. 2010. EU agricultural production and trade: can more efficiency prevent increasing 'land grabbing' outside of Europe? OPERA Research Center. www.appgagscience.org.uk/linkedfiles/Final_Report_Opera.pdf (accessed June 20, 2013).

World Bank. 2009. *The World Bank Annual Report 2009*. Washington, DC: World Bank.

World Bank. 2009. http://databank.worldbank.org/data/ (accessed March 20, 2019).

World Future Council & IFOAM. 2018. *Future Policy Award 2018. Scaling up Agroecology. Evaluation Report*. Bonn, Hamburg, Geneva: World Future Council & IFOAM–Organics International.

World Resources Institute. 1999. *La Situación Del Mundo, 1999*. Lester Brown, Christopher Flavin & Hillary French (eds.). Barcelona: Icaria.

World Resources Institute. 2002. *La Situación Del Mundo, 2002*. Lester Brown, Christopher Flavin & Hillary French (eds.). Barcelona: Icaria.

Wright, I. 2005. The social architecture of capitalism. *Physica A*, 346, 589–620.

Wright, I. 2017. The social architecture of capitalism. Things are getting worse. *From Here to There. Adventures in Marxist Theory*. November 17, 2017. (https://ianwrightsite.wordpress.com/2017/11/16/the-social-architecture-of-capitalism/(accessed March 2, 2019).

Wrigley, E.A. 1985. *Historia y Población. Introducción a la demografía histórica*. Barcelona: Crítica.

Wrigley, E.A. 1993. *Cambio, continuidad y azar. Carácter de la Revolución Industrial inglesa*. Barcelona: Editorial Crítica.

Wrigley, E.A. 2004. *Poverty, Progress, and Population*. Cambridge: Cambridge University Press.

Zanias, G.P. 2005. Testing for trends in the terms of trade between primary commodities and manufactured goods. *Journal of Development Economics*, 78, 49–59.

Zhou, Y. 2000. Smallholder Agriculture, Sustainability and the Syngenta Foundation. Syngenta Foundation. https://pdfs.semanticscholar.org/b6b9/3f6cdeffc8b92278df329c4a2662b80a1bbb.pdf.

Index

Page numbers in *italic* denote figures and those in **bold** denote tables. Footnotes are denoted by a letter n between page number and note number.

Hayek, F. 108
Healthy diets 114, 135–138, 154
Heller, M.C 41
Herbicide resistance 63
Hobbes, Thomas 92, 94
Hobsbawn, E. 101
Holloway, J. 98–99
Ho, Mae-Wan 19
Hornborg, Alf 30
Household energy consumption **42**, 43
Human right to food 55, 137
Hunger 55, 137

Ideology for agroecological transition
    73–76, 79–80
IFOAM-Organics International 147, 148, 164
Import substitution policy 35
Incomes, farm 38, 45, 54–55, 60, 64, *64*
India, agroecological policies/practices 115,
    163–164
Industrial agriculture crisis 54–71
    characterization of 54–60
    evidences of collapse risk 60–65
    food complex 69–71
    proposed alternatives 65–69
Industrialization of agriculture 21, 27–39
    first food regime 33–34
    Green Revolution 34–37, **35**
    livestock industry 30–31, 32, 35, 36, 42, 45, 58,
        63–65, 70
    main drivers of 37–39
    origins of 28–31
    peasant conditions under 128–131, *130*
    private property and market 31–33
    production intensification 34, 35, 37–39, 59, 62
    productive specialization 30, 33–34,
        37–39, 59, 62
    second food regime 34–37, **35**
    third food regime 44–45
Industrialization of food system 40–45, **42**
Information campaigns 156
Information flows 4–5, 10–11, 21–23
Institutional design for agroecological transition
    73, 74–75, 78–96; see also Public policies
    favoring Agroecology (PPfA)
    agroecological districts 88, 158
    agroecological effects of cooperative design 79
    community currencies 88–89
    cooperative democracy 93–95
    cooperative institutions as preferential
        model 84
    cooperative resource management principles
        74, 78–79
    deliberative democracy 95–96
    democratic governance 89–96, 154

diversity of institutions 83–89
family institutions 83–84
local markets 85–86
multilevel collective action 90–92
normative action 92–93
popular sovereignty 92–93
reforming global trade 86–88
resilience of institutions 80–81
scales and "social point" of cooperative
    institutions 82–83
Intensification
    ecological 68–69
    industrial agriculture 34, 35, 37–39, 59, 62
    sustainable 67–68
Intensive livestock model 63–65
Intergenerational solidarity effect 79
International Covenant on Economic, Social and
    Cultural Rights (ICESCR) 137
Investment products 71
IPES-Food report 71, 111
Irrigated agriculture 57, **57**, 59

Jevons Paradox 52

Keoleian, G.A. 41
Knowledge
    as common good 110
    peasant 126
Krausmann, F. 29, 30, 33, 36, 51

Lachman, D.A. 25
Laclau, Ernesto 120–121
Laforge, J.M.L. 112
Livestock industry 30–31, 32, 35, 36, 42, 45, 58,
    63–65, 70
Lobbying 71
Local currencies 88–89
Localization effect 79
Local markets 85–86
Lock-ins 69, 111
Long, N. 108
Loos, J. 69
Lowder, S.K. 60–62
Luhmann, Niklas 5, 8
Luxemburg, Rosa 98

McMichael, P. 45
McNeill, John 97
Malnutrition 55, 137
Marcus J. 12
Markets 22, 23–24, 30, 31–33, 34, 50,
    106–109, 128
    local 85–86
    territorial 110
Marxist theory 23, 51–54, 120, 138–139, 145